Number Patterns and Sequences: Basics of Mathematical Patterns

NUMBER PATTERNS AND SEQUENCES: BASICS OF MATHEMATICAL PATTERNS

Prof. Alok Kumar Verma

ARCLER
PRESS

www.arclerpress.com

Number Patterns and Sequences: Basics of Mathematical Patterns

Prof. Alok Kumar Verma

Arcler Press
224 Shoreacres Road
Burlington, ON L7L 2H2
Canada
www.arclerpress.com
Email: orders@arclereducation.com

ISBN: 9781-77469-944-7 (Hardcover)

ABOUT THE AUTHOR

Dr. Alok Kumar Verma is an accomplished mathematician with an impressive academic background, holding an M.Sc. in Mathematics, M.Tech, and a Ph.D. in Mathematics, which he was awarded by Kumaun University Nainital on December 7, 2006. With a wealth of experience spanning over twenty-three years, he currently serves as a distinguished Professor and Head of the Department of Mathematical Sciences and Computer Applications. Dr. Verma has made significant contributions to the field of mathematics, with 13 research papers published in esteemed international and national journals, as well as authoring two books. He is also an active participant in conferences and workshops, where he presents his research findings. Additionally, Dr. Verma's commitment to academia is evident through his involvement in various academic committees and organizations, such as being a member of the grants committee at Raja Ram Mohan Roy Library Foundation in Kolkata and the Executive Council of Bundelkhand University Jhansi in multiple terms, among other prestigious affiliations. His dedication and contributions continue to make a positive impact on the academic community.

TABLE OF CONTENTS

LIST OF FIGURES

LIST OF ABBREVIATIONS

CPI	Consumer Price Index
ECC	Elliptic Curve Cryptography
ECDLP	Elliptic Curve Discrete Logarithm Problem
GNFS	General Number Field Sieve
IFS	Iterated Function Systems
MACD	Moving Average Convergence Divergence
PMF	Probability Mass Function
PRNGs	Pseudorandom Number Generators
RSI	Relative Strength Index

PREFACE

Number patterns are recurring sequences of numbers that exhibit consistent relationships based on mathematical operations and properties. Number patterns are subjected to rigorous analysis within the field of mathematics, with each pattern exhibiting distinct properties and behaviors. These patterns serve as abstract representations of numerical structures, often characterized by regular increments, multiplicative factors, or iterative processes. The study of number patterns falls within the domain of discrete mathematics, encompassing topics such as arithmetic progressions, geometric progressions, and more complex recursive sequences. These patterns are elucidated through algebraic expressions and equations, enabling mathematicians to deduce specific terms within a sequence based on established formulas. Number patterns provide foundational insights into the behavior of numbers, serving as essential tools in various mathematical analyses and applications.

In the world of mathematical investigation, "Number Patterns and Sequences: Basics of Mathematical Patterns" stands as a gateway to the intricate symphony woven by numbers. This book embarks on a profound journey through the fascinating world of numerical relationships, with each chapter serving as a meticulously crafted lens, revealing the underlying elegance and significance of diverse number patterns and sequences.

Chapter 1 introduces the concept of number patterns. This chapter serves as a compass, guiding us through the landscape of mathematical relationships that characterize the fabric of numbers. Chapter 2 unfurls the rich narratives woven by arithmetic and geometric sequences. Here, the interplay between these sequences is unveiled. Chapter 3 opens the door to the enchanting world of the Fibonacci sequence. Revealing the profound connection between mathematics and the natural world. Chapter 4 unravels the mysteries of prime numbers.

Chapter 5 introduces the mathematical tapestry of Pascal's Triangle and Binomial Coefficients. In Chapter 6, the concept of recurrence relations and recursive sequences takes center stage. Chapter 7 introduces us to the realm of Lucas Numbers and Higher-Order Sequences. As companions to the Fibonacci sequence, these numbers unveil their own tales of growth and connection, broadening our understanding of the intricacies inherent in mathematical patterns.

Chapter 8 transports us into the realm of advanced number patterns. Here, the journey reaches its apex, unveiling the extraordinary patterns that emerge when mathematical concepts are pushed to their limits.

The culmination of collaborative efforts, this book is a testament to the dedication of esteemed mathematicians. Each chapter is a product of their expertise, inviting readers on a captivating journey through the realms of numerical exploration. It is our hope that this book serves as a guiding light, inspiring further exploration, contemplation, and a deeper appreciation for the intricate dance of numbers.

CHAPTER 1

Introduction to Number Patterns

Contents

1.1. INTRODUCTION

A number pattern is a sequence of numbers that follows a specific rule. The rule can be simple, such as adding 1 to each number to get the next number, or it can be more complex. Also, a number sequence is a list of numbers that are in order. The order of the numbers in a sequence is important because it determines the pattern (Yesildere & Akkoc, 2012).

Number patterns are important in mathematics and other fields. In mathematics, they can be used to solve equations and inequalities, model the growth of populations, and predict future events.

In other fields, such as science, engineering, and finance, number patterns can be used to design bridges, predict stock prices, and much more (Liljedahl, 2004). There are many different types of number patterns. Some of the most common types include:

1. **Arithmetic Sequences**: In an arithmetic sequence, each number is equal to the previous number plus a constant value. For example, the sequence 1, 4, 7, 10, 13 is an arithmetic sequence because each number is 3 more than the previous number.
2. **Geometric Sequences:** In a geometric sequence, each number is equal to the previous number times a constant value. For example, the sequence 2, 4, 8, 16, 32 is a geometric sequence because each number is 2 times the previous number.
3. **Quadratic Sequences:** In a quadratic sequence, each number is equal to the square of a number (x) with the addition of that number plus a constant value. For example, the sequence $x^2 + x + 1$ is a quadratic sequence.
4. **Fibonacci Sequences:** Fibonacci sequences/mm/ are a special type of number sequence where each number is equal to the sum of the two previous numbers. The Fibonacci sequence looks like 0, 1, 1, 2, 3, 5, 8, 13, 21, 34, … As the Fibonacci sequence advances, the relationship between consecutive numbers within the sequence progressively approaches a constant value known as Phi (1.6180…), denoted as Φ, the 21st letter of the Greek alphabet. Phi is also renowned as the Golden Ratio.

There are a few ways to identify number patterns. One way is to look for common differences or ratios. For example, if you notice that each number in a sequence is C (any constant) more than the previous number, then you can conclude that the sequence is an arithmetic sequence (Zazkis & Campbell,

2006). Another way to identify number patterns is to write down the first few terms of the sequence. Once you have written down a few terms, you can try to find a formula for the terms of the sequence. For example, the formula for the arithmetic sequence 1, 4, 7, 10, 13 is $(a_n = n + 3)$, where a_n is the nth term in the sequence (Zippert et al., 2021). Number patterns can also be used in scientific and statistical phenomena. For example, the Fibonacci sequence is often used to model the growth of populations because it has a self-similar structure that is similar to the growth of many populations. In engineering, number patterns can be used to design bridges. For instance, the Golden Ratio is often used in the design of bridges because it is a very pleasing proportion that is also found in nature. In finance, number patterns can be used to predict stock prices. The Elliott Wave Theory is a popular theory that uses number patterns to predict the future direction of stock prices (Wakefield, 2000).

1.2. TYPES OF NUMBER PATTERNS

Following are the types of the number patterns:

1.2.1. Arithmetic Sequences

An arithmctic sequence is a special type of linear pattern in which the common difference is always positive. This means that each term is greater than the previous term. For example, sequence 2, 5, 8, 11, 14. is an arithmetic sequence because each term is 3 more than the previous term. The common difference in this sequence is 3 (Herbert & Brown, 1997).

The formula for an arithmetic sequence is the same as the formula for a linear pattern:

$$\boldsymbol{a_n = a_1 + d(n-1)}$$

Where n is the number of terms, a_n is the nth term in the sequence, a_1 is the first term in the sequence, and d is the common difference. Whereas the common difference d is always positive (Rathgeb & Green, 2013).

1.2.2. Geometric Sequences

A geometric sequence is a special type of linear pattern in which the common difference is always multiplied by a constant value. This means that each term is equal to the previous term times a constant value.

The formula for a geometric sequence is:

$$a_n = a_1 r^{n-1}$$

where a_n is the nth term in the sequence, a_1 is the first term in the sequence, and is the common ratio.

The common ratio in a geometric sequence can be positive or negative. If the common ratio is positive, the sequence will grow exponentially. If the common ratio is negative, the sequence will decay exponentially (Cover, 1965).

1.2.3. Quadratic Patterns

A quadratic pattern is a sequence of numbers in which the difference between any two consecutive terms is not constant. This means that the second difference between any two consecutive terms is constant. For example, sequence 1, 2, 4, 7, 11 is a quadratic pattern. The second difference between any two consecutive terms is 1 because 4 – 2 = 2, 7 – 4 = 3, and 11 – 7 = 4 (Field & Golubitsky, 2009).

Quadratic patterns can be represented by the following formula:

$$a_n = a_1 + b(n-1) + c(n-1)^2$$

where a_n is the nth term in the sequence, a_1 is the first term in the sequence, b is the linear coefficient, and c is the quadratic coefficient. The linear coefficient b controls the slope of the sequence, and the quadratic coefficient c controls the curvature of the sequence (Resnik, 1981).

The Figure 1.1 shows a quadratic pattern with the first term 1, the linear coefficient 2, and the quadratic coefficient 1. As you can see, the sequence starts off increasing slowly, then increases more rapidly, and then levels off.

Figure 1.1. Illustration of the quadratic pattern.

Source: chrishunter.ca, Creative Commons License.

Quadratic patterns can be found in many different areas, such as physics, chemistry, biology, and economics. For example, in physics, quadratic patterns can be used to model the motion of objects under the influence of gravity. In chemistry, quadratic patterns can be used to model the growth of chemical reactions. In biology, quadratic patterns can be used to model the spread of diseases. In economics, quadratic patterns can be used to model the growth of economies.

1.2.4. Other Types of Patterns

1.2.4.1. Trigonometric Patterns

Trigonometric patterns are sequences of numbers that follow a trigonometric function, such as sine or cosine. For example, the sequence $sin\ (n\varpi)$ is a trigonometric pattern.

1.2.4.2. Exponential Patterns

Exponential patterns are sequences of numbers that grow or decay exponentially. For example, the sequence 2^x is an exponential pattern (Coe & Ruthven, 1994) (Figure 1.2).

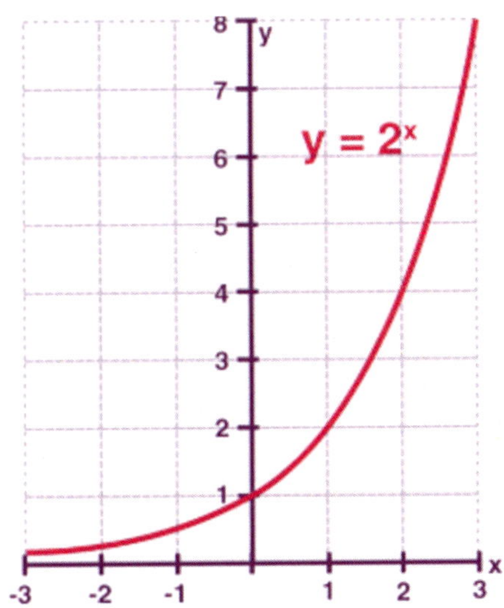

Figure 1.2. Graphical representation of exponential function.

Source: Byjus, commons license.

1.2.4.3. Logarithmic Patterns

Logarithmic patterns are sequences of numbers that exhibit logarithmic growth or are generated using logarithmic functions. For example, the sequence $y = \log_2(x)$ is an example of a logarithmic pattern (Figure 1.3).

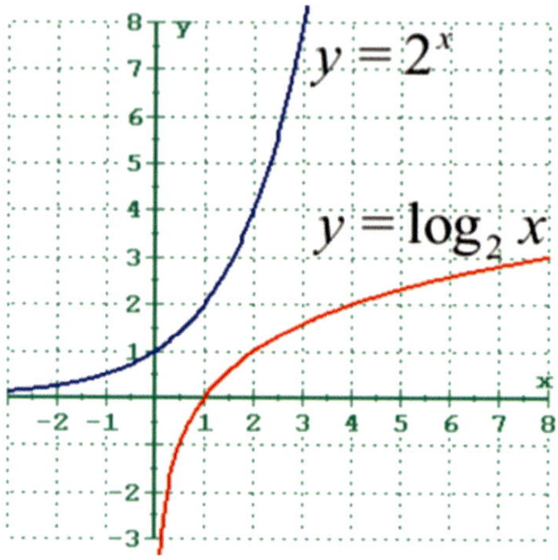

Figure 1.3. Comparison of logarithmic and exponential patterns.

Source: People.richland.edu, Creative Commons License.

1.2.4.4. Fibonacci Patterns

Fibonacci patterns are sequences of numbers where each number is equal to the sum of the two previous numbers (Artstein et al., 2014). In mathematics, the Fibonacci sequence denoted as "Fn" is established through a recursive formula, initialized with the starting values F0=0 and F1=1, expressed as (Figure 1.4):

$$F_n = F_{n-1} + F_{n-2}$$

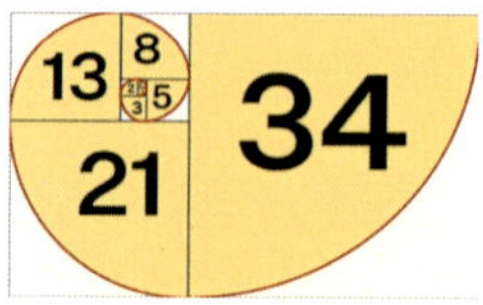

Figure 1.4. Illustration of Fibonacci pattern.

Source: Cuemath, Creative Commons License.

1.2.4.5. Pascal's Triangle

Pascal's triangle is a triangular array of numbers where each number is the sum of the two numbers above it (Figure 1.5).

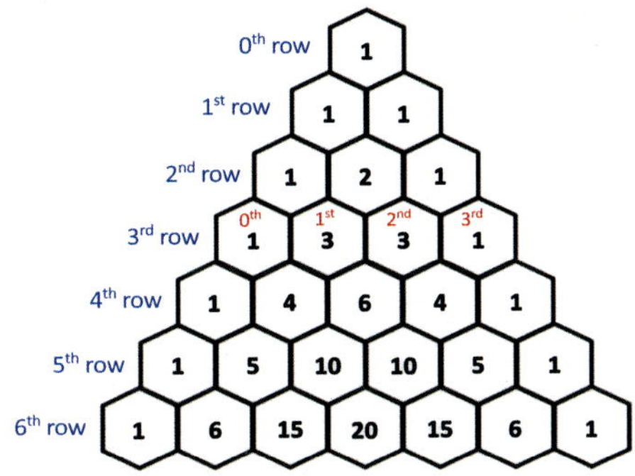

Figure 1.5. Illustration of a Pascal's triangle.

Brilliant Org, creative commons.

1.3. IDENTIFICATION OF NUMBER PATTERNS

Number patterns can be identified by looking for common differences (a constant value added to each term in a sequence to get the next term) or ratios between consecutive terms. Also, the number patterns can be identified by writing the first few terms of the sequence and by the formulas of the number pattern.

Once you have identified a number pattern, you can use it to solve problems or make predictions. For example, if you know that a sequence is an arithmetic sequence, you can use the formula to find the next term in the sequence (Herman et al., 1973).

Following is the list of techniques for identifying number patterns (Reed, 1997):

i. Look for patterns in the first few terms of the sequence.
ii. Try to find a formula that generates the terms of the sequence.
iii. Look for common differences or ratios between consecutive terms.
iv. Use a graph to visualize the sequence.

1.3.1. Common Differences and Ratios

The use of common difference is the most common method in the identification of the number pattern. For example, the sequence 1, 4, 7, 10, 13 is an arithmetic sequence with a common difference of 3. This means that we can get from any term in the sequence to the next term by adding 3. Also, we can find the common difference of a sequence by subtracting any two consecutive terms. For example, in the sequence 1, 4, 7, 10, 13, the common difference is 4 – 1 = 3 (Bickley, 1995). A common ratio is a constant value that is multiplied by each term in a sequence to get the next term. For example, sequence 2, 4, 8, 16, 32, … is a geometric sequence with a common ratio of 2. This means that we can get from any term in the sequence to the next term by multiplying by 2. We can find the common ratio of a sequence by dividing any two consecutive terms. For example, in 2, 4, 8, 16, 32, …, the common ratio is 4 / 2 = 2 (Zazkis & Lijedahl, 2002).

1.3.2. Writing Down the First Few Terms of the Pattern

One way to identify number patterns is to write down the first few terms of the pattern. For example, sequence 1, 4, 7, 10, 13,… is an arithmetic sequence because the first few terms are all consecutive odd numbers. The sequence 2, 4, 8, 16, 32 is a geometric sequence because the first few terms are all powers of 2 (Sun, 1996).

Tip

When trying to identify number patterns, it's often helpful to write down the first few terms of the pattern. This can provide visual clues to its nature.

We can also use the first few terms of a pattern to find a formula that generates the terms of the pattern. For example, if we know that the sequence is an arithmetic sequence with a common difference of 3 and the first term is 1, then we can find the next term by adding 3 to the first term, which gives us 4.

1.3.3. Formulas for Number Patterns

There are many different formulas that can be used to generate the terms of a number pattern. Some of the most common formulas that have already been discussed are arithmetic sequence formula, geometric sequence formula, and quadratic pattern formula.

We can identify number patterns by trying to find a formula for the terms of the pattern. If we can find a formula that generates all the terms in the sequence, then we can be sure that the sequence is a number pattern (Granville & Soundararajan, 2007). Also, we can use formulas to find the next term in a sequence. For example, if we know that the sequence is an arithmetic sequence with a common difference of 3, then we can find the next term by adding 3 to the current term.

1.4. APPLICATIONS OF NUMBER PATTERNS

Number patterns have numerous practical applications across various fields. In finance, they are employed for calculating interest and predicting future financial trends.

In science and engineering, they help model natural phenomena, for example, the behavior of physical systems. In computer science, number patterns are crucial for algorithms, data compression, and cryptography (Nogami et al., 2014). Additionally, in education, they serve as a fundamental concept for teaching mathematical concepts and problem-solving skills. Number patterns play a fundamental role in understanding and addressing real-world problems, making them a valuable tool in many aspects of modern life.

1.4.1. Mathematics

1.4.1.1. Solving Equations

Number patterns can be used to solve equations by finding a formula for the terms of the pattern. For example, the equation $x+2=5$ can be solved by writing the left-hand side of the equation as an arithmetic sequence with a common difference of 2 (Radicic, 2016). The solution to the equation is the first term of the sequence.

1.4.1.2. Other Applications

Number patterns can also be applied to solve various types of problems in mathematics, including differential equations and optimization problems.

For instance, certain differential equations, like $y' = y$, have solutions that can be represented as geometric sequences.

The applications of number patterns in mathematics are diverse and valuable for problem-solving, helping individuals develop their mathematical skills and gain a deeper understanding of the world around them (Potter, 1972).

Here are some other specific examples of the applications of number patterns in solving equations and inequalities:

The Fibonacci sequence is utilized to solve the Fibonacci recurrence relation:

1. The Fibonacci sequence is employed to solve the Fibonacci recurrence relation:

   $x_n = x_{n-1} + x_{n-2}$

2. The Golden ratio is utilized to address the Babylonian reciprocal equation:

   $x_1 + y_1 = 1$

3. The quadratic formula is applied to solve quadratic equations of the form:

   $ax^2 + bx + c = 0$

1.4.2. Statistics

1.4.2.1. Modeling the Growth of Populations

The growth of populations can be modeled by using an arithmetic sequence. The first term in the sequence is the initial population size, and the common difference is the growth rate. For example, if the initial population size is 100 and the growth rate is 2%, then the sequence of population sizes would be 100, 102, 104.4, 106.84, and so on (Vogel, 2005).

1.4.2.2. Modeling the Spread of Diseases

The spread of diseases can be modeled by using a geometric sequence. The first term in the sequence is the number of infected individuals, and the common ratio is the transmission rate.

For example, if the initial number of infected individuals is 1 and the transmission rate is 0.5, then the sequence of number of infected individuals would be 1, 0.5, 0.25, 0.125, and so on. In both cases, the number patterns can be used to make predictions about the future growth of the population or the spread of the disease. For example, we can use the number patterns to predict when the population will reach a certain size or when the disease will infect a certain number of people (Reinholz & Shah, 2018).

1.4.3. Science

Here are some other examples of how number patterns are used in science:

1. To model the movement of objects under the influence of gravity.
2. To predict the behavior of waves.
3. To pattern the growth of crystals.
4. To simulate the evolution of species.

1.4.4. Engineering

Engineering is the application of science and mathematics to solve practical problems. Number patterns can be used in engineering to design bridges, buildings, and other structures (Azaryahu et al., 2023).

Here are some specific examples of how number patterns are used in engineering:

1. In designing buildings, engineers use number patterns to calculate the stresses and strains on the building. They also use number patterns to design the shape of the building to make it as strong as possible.
2. In designing machines, engineers use number patterns to design the gears, linkages, and other components of machines. They also use number patterns to calculate the forces and torques on these components.
3. In designing electrical circuits, engineers use number patterns to calculate the currents, voltages, and resistances in electrical circuits. They also use number patterns to design the components of electrical circuits.

The use of number patterns in engineering is essential to ensure the safety and reliability of structures and machines (Yerushalmy, 2000).

1.4.5. Predicting Stock Prices

Stock prices can be modeled by a variety of mathematical patterns, including arithmetic sequences, geometric sequences, and quadratic sequences. By identifying the pattern that best fits the historical data, we can make predictions about future stock prices. For example, the daily stock price of a company might be modeled by an arithmetic sequence with a common difference of 10. This means that the stock price will increase by 10 each day. If the current stock price is $100, then we can predict that the stock price will be $110 on the next day and $120 after two days (Burch et al., 2017).

Did You Know?

Number patterns play a role in predicting stock prices, as certain mathematical models and algorithms are used to analyze historical price data and identify trends.

1.4.5.1. Predicting Interest Rates

Interest rates can also be modeled by mathematical patterns. For example, annual interest rates might be calculated by a geometric sequence with a common ratio of 1.05. This means that the interest rate will increase by 5% each year. If the current interest rate is 5%, then we can predict that the interest rate will be 5.25% next year and 5.51% the year after that (Zimmermann & Cunningham, 1991).

1.4.5.2. Other Applications in Finance

Number patterns can also be used to predict other financial data, such as the exchange rate between two currencies, the price of a commodity, and the value of a bond. By understanding the patterns that drive these financial data, we can make better predictions about the future. Here are some other specific examples of the applications of number patterns in finance:

1. **Elliott Wave Theory:** The Elliott Wave Theory is a popular theory that uses Fibonacci sequences to predict stock prices.
2. **Moving Average Convergence Divergence (MACD):** The MACD is a technical indicator that uses moving averages to identify trends and reversals in stock prices.
3. **Relative Strength Index (RSI):** The RSI is a technical indicator that measures the momentum of a stock price.

These are just a few examples of the many ways that number patterns can be used in finance.

1.4.6. Other Fields

1.4.6.1. Music

Number patterns are used in music to create melodies, harmonies, and rhythms. For example, the Fibonacci sequence is used in the construction of many musical pieces, such as Bach's "Well-Tempered Clavier." The Fibonacci sequence is also used in the construction of musical instruments, such as the violin and the piano. In music, number patterns can be used to create a sense of order and beauty. By understanding the mathematical principles behind number patterns, musicians can create music that is both pleasant and meaningful (Azaryahu et al., 2023).

1.4.6.2. Art

Number patterns are used in art to create symmetry, proportion, and balance. For example, the golden ratio is used in the construction of many paintings, sculptures, and buildings. The golden ratio is also used in the design of many everyday objects, such as credit cards and computer screens. In art, number patterns can be used to create a sense of harmony and beauty. By understanding the mathematical principles behind number patterns, artists can create art that is both attractive and thought-provoking.

1.4.6.3. Architecture

Number patterns are used in architecture to create structures. For example, the Fibonacci sequence is used in the construction of many buildings, such as the Parthenon and the Sagrada Familia. Number patterns can be used to create a sense of order. By understanding the mathematical principles behind number patterns, architects can create structures that are both strong and timeless.

Activity 1.1

Imagine you are a data scientist analyzing historical stock price data. You notice a recurring number pattern in the stock prices of a particular company. Describe the steps you would take to analyze this pattern and make predictions about future stock prices based on it.

1.5. SUMMARY

This chapter introduces number patterns, elucidating their relevance in mathematics and various applications. It classifies them into three fundamental categories: Arithmetic Sequences (with a constant difference between consecutive terms), Geometric Sequences (with a constant ratio between terms), and Quadratic Patterns (exhibiting a quadratic growth function). The chapter explores additional patterns as well. It emphasizes the identification of patterns using common differences, ratios, and the initial term identification process. Formulas for describing number patterns are introduced. The chapter underscores practical applications, including mathematical problem-solving, scientific modeling, engineering, and even financial prediction. The goal is to provide a comprehensive understanding of number patterns, their identification, formulas, and real-world relevance.

REFERENCES

1. Amit, D. J., Gutfreund, H., & Sompolinsky, H., (1985). Storing infinite numbers of patterns in a spin-glass model of neural networks. *Physical Review Letters, 55*(14), 1530.
2. Artstein, Z., Borwein, J., Van, D. P. A., Shallit, J., Zudilin, W., Franklin, J., & Prince, G., (2014). *Mathematics and the Real World, 1*(1), 1–20.
3. Azaryahu, L., Broza, O., Cohen, S., Hershkovitz, S., & Adi-Japha, E., (2023). Development of creative thinking patterns via math and music. *Thinking Skills and Creativity, 47*, 101196.
4. Becker, J. R., & Rivera, F., (2006). Sixth graders' figural and numerical strategies for generalizing patterns in algebra. In: *Proceeding of the 28th Annual Meeting of the North American Chapter of the International Group for the Psychology of Mathematics Education* (Vol. 2, pp. 95–101).
5. Bickley-Green, C. A., (1995). Math and art curriculum integration: A post-modern foundation. *Studies in Art Education, 37*(1), 6–18.
6. Burch, M., Hlawatsch, M., & Weiskopf, D., (2017). Visualizing a sequence of a thousand graphs (or even more). In: *Computer Graphics Forum* (Vol. 36, No. 3, pp. 261–271).
7. Coe, R., & Ruthven, K., (1994). Proof practices and constructs of advanced mathematics students. *British Educational Research Journal, 20*(1), 41–53.
8. Cover, T. M., (1965). Geometrical and statistical properties of systems of linear inequalities with applications in pattern recognition. *IEEE Transactions on Electronic Computers*, (3), 326–334.
9. Dörfler, W., (2008). En route from patterns to algebra: Comments and reflections. *ZDM*, *40*, 143–160.
10. Falcon, S., & Plaza, Á., (2007). The k-Fibonacci sequence and the Pascal 2-triangle. *Chaos, Solitons & Fractals, 33*(1), 38–49.
11. Field, M., & Golubitsky, M., (2009). *Symmetry in Chaos: A Search for Pattern in Mathematics, Art, and Nature* (Vol. 1, No. 1, pp. 1–20). Society for Industrial and Applied Mathematics.
12. Gamoran, A., & Hannigan, E. C., (2000). Algebra for everyone? Benefits of college-preparatory mathematics for students with diverse abilities in early secondary school. *Educational Evaluation and Policy Analysis, 22*(3), 241–254.

13. Granville, A., & Soundararajan, K., (2007). An uncertainty principle for arithmetic sequences. *Annals of Mathematics, 1*(1), 593–635.
14. Herbert, K., & Brown, R. H., (1997). Patterns as tools for algebraic reasoning. *Teaching Children Mathematics, 3*(6), 340–344.
15. Herman, G. T., Lent, A., & Rowland, S. W., (1973). ART: Mathematics and applications: A report on the mathematical foundations and on the applicability to real data of the algebraic reconstruction techniques. *Journal of Theoretical Biology, 42*(1), 1–32.
16. Hildenbrand, D., (2012). Foundations of geometric algebra computing. In: *AIP Conference Proceedings* (Vol. 1479, No. 1, pp. 27–30).
17. Lannin, J. K., Barker, D. D., & Townsend, B. E., (2006). Recursive and explicit rules: How can we build student algebraic understanding? *The Journal of Mathematical Behavior, 25*(4), 299–317.
18. Liljedahl, P., (2004). Repeating pattern or number pattern: The distinction is blurred. *Focus on Learning Problems in Mathematics, 26*(3), 24–42.
19. MacEachren, A. M., & Ganter, J. H., (1990). A pattern identification approach to cartographic visualization. *Cartographica: The International Journal for Geographic Information and Geovisualization, 27*(2), 64–81.
20. Morse, J., (1999). Recursion and explicit formulas for particular n-variable Knop–Sahi and Macdonald polynomials. *Journal of Combinatorial Theory, Series A*, *88*(1), 1–18.
21. Mukherjee, A., Mukhi, S., & Pakman, A., (2007). FZZ algebra. *Journal of High Energy Physics, 2007*(1), 025.
22. Neuts, M. F., (1977). Some explicit formulas for the steady-state behavior of the queue with semi-Markovian service times. *Advances in Applied Probability, 9*(1), 141–157.
23. Nogami, Y., Tada, K., & Uehara, S., (2014). A geometric sequence binarized with Legendre symbol over odd characteristic field and its properties. *IEICE Transactions on Fundamentals of Electronics, Communications and Computer Sciences, 97*(12), 2336–2342.
24. Oevel, W., & Fuchssteiner, B., (1982). Explicit formulas for symmetries and conservation laws of the Kadomtsev-Petviashvili equation. *Physics Letters A*, *88*(7), 323–327.
25. Pegg, J., & Redden, E., (1990). From number patterns to algebra: The important link. *Australian Mathematics Teacher, 46*(2), 19–22.

26. Potter, D. C. D., (1972). Triangular square numbers. *The Mathematical Gazette, 56*(396), 109–110.
27. Puig, L., & Rojano, T., (2004). The history of algebra in mathematics education. *The Future of the Teaching and Learning of Algebra The 12th ICMI Study, 1*(1), 187–223.
28. Radičić, B., (2016). On k-circulant matrices (with geometric sequence). *Quaestiones Mathematicae, 39*(1), 135–144.
29. Rathgeb-Schnierer, E., & Green, M., (2013). Flexibility in mental calculation in elementary students from different math classes. In: *Proceedings of the Eighth Congress of the European Society for Research in Mathematics Education* (Vol. 1, No. 1, pp. 353–362).
30. Reed, M., (1997). Algebraic structure of genetic inheritance. *Bulletin of the American Mathematical Society, 34*(2), 107–130.
31. Reinholz, D. L., & Shah, N., (2018). Equity analytics: A methodological approach for quantifying participation patterns in mathematics classroom discourse. *Journal for Research in Mathematics Education, 49*(2), 140–177.
32. Resnik, M. D., (1981). Mathematics as a science of patterns: Ontology and reference. *Nous, 1*(1), 529–550.
33. Rivera, F. D., & Becker, J. R., (2007). Abduction–induction (generalization) processes of elementary majors on figural patterns in algebra. *The Journal of Mathematical Behavior, 26*(2), 140–155.
34. Silva, C. M., Moses, R. P., Rivers, J., & Johnson, P., (1990). The algebra project: Making middle school mathematics count. *The Journal of Negro Education, 59*(3), 375–391.
35. Sudakov, I., Bellsky, T., Usenyuk, S., & Polyakova, V., (2014). Mathematics and climate infographics: A mechanism for interdisciplinary collaboration in the classroom. *PRIMUS, 26*(2), 158–167.
36. Sun, G. Q., Jusup, M., Jin, Z., Wang, Y., & Wang, Z., (2016). Pattern transitions in spatial epidemics: Mechanisms and emergent properties. *Physics of Life Reviews, 19*, 43–73.
37. Sun, Z. W., (1996). Covering the integers by arithmetic sequences. II. *Transactions of the American Mathematical Society, 348*(11), 4279–4320.
38. Ulam, S., (1962). On some mathematical problems connected with patterns of growth of figures. In: *Proceedings of Symposia in Applied*

Mathematics (Vol. 14, No. 14, pp. 215–224). Providence, RI, USA: American Mathematical Society.

39. Vogel, R., (2005). Patterns—A fundamental idea of mathematical thinking and learning. *ZDM, 37*, 445–449.
40. Wakefield, D. V., (2000). Math as a second language. In: *The Educational Forum* (Vol. 64, No. 3, pp. 272–279).
41. Yerushalmy, M., (2000). Problem solving strategies and mathematical resources: A longitudinal view on problem solving in a function-based approach to algebra. *Educational Studies in Mathematics, 43*(2), 125–147.
42. Yeşildere, İ. S., & Akkoç, H., (2012). Investigating the development of prospective mathematics teachers' pedagogical content knowledge of generalizing number patterns through school practicum. *Journal of Mathematics Teacher Education, 15*, 207–226.
43. Zazkis, R., & Campbell, S. R., (2006). *Number Theory in Mathematics Education: Perspectives and Prospects, 1*(1), 1–20.
44. Zazkis, R., & Liljedahl, P., (2002). Arithmetic sequence as a bridge between conceptual fields. *Canadian Journal of Science, Mathematics and Technology Education, 2*, 91–118.
45. Zimmermann, W., & Cunningham, S., (1991). Editor's introduction: What is mathematical visualization. *Visualization in Teaching and Learning Mathematics, 1*(8), 1–20.
46. Zippert, E. L., Douglas, A. A., Tian, F., & Rittle-Johnson, B., (2021). Helping preschoolers learn math: The impact of emphasizing the patterns in objects and numbers. *Journal of Educational Psychology, 113*(7), 137.

CHAPTER 2

Arithmetic and Geometric Sequences

Contents

2.1. INTRODUCTION

In the domain of mathematical patterns and sequences, two fundamental structures stand as bases of numerical organization: the Arithmetic Sequence and the Geometric Sequence. These precise constructs, deeply rooted in mathematics, serve as the building blocks for numerous mathematical and real-world phenomena (Rahmani-Andebili, 2021).

Arithmetic sequences provide a foundational understanding of how numbers evolve in a predictable and consistent manner. By encapsulating the concept of a common difference between consecutive terms, arithmetic sequences teach us to recognize and work with the fundamental idea of linear growth.

This understanding extends beyond mathematics, influencing comprehension of rates of change and gradual progression in various aspects of the physical world.

Furthermore, arithmetic sequences serve as a gateway to more advanced mathematical concepts, such as calculus, where they underpin the understanding of limits and derivatives. In the practical realm, they find applications in fields like finance and engineering, aiding in the modeling of linearly changing quantities.

As an educational tool, arithmetic sequences help learners develop problem-solving skills and mathematical reasoning, making them an essential component of mathematical education (Rahmani-Andebili, 2021) (Figure 2.1).

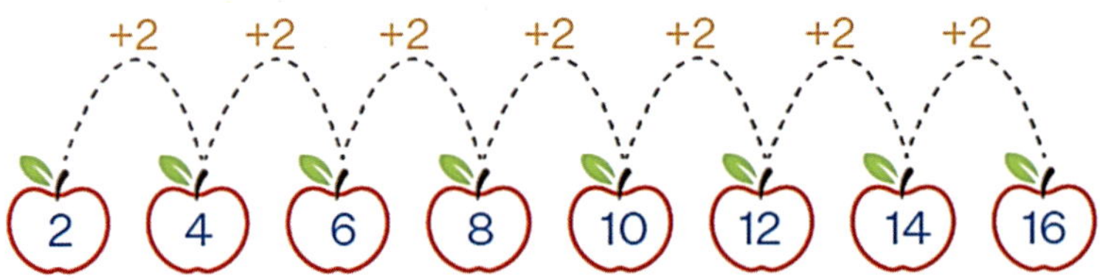

Figure 2.1. Illustration of an arithmetic sequence.

Source: Cuemath, Creative Commons License.

Geometric sequences offer a unique perspective on the progression of numbers, demonstrating the power of exponential growth or decay. The common ratio between consecutive terms in a geometric sequence highlights the concept of multiplication as the driving force behind this growth. In mathematics, geometric sequences provide insights into exponential functions and the concept of limits, which are crucial in calculus. Moreover, they play a vital role in finance, modeling scenarios like compound interest and exponential growth in investments. In physics and engineering, geometric sequences help describe processes with exponential behavior, such as radioactive decay.

In computer science, they are essential in algorithms, particularly those involving exponential time complexity. Beyond these applications, geometric sequences foster a deep understanding of exponential phenomena and serve as a fundamental educational tool, helping students grasp exponential functions, geometric series, and the significance of exponential growth in various real-world contexts (Camouzis & Kotsios, 2021).

2.2. ARITHMETIC SEQUENCES

In mathematical sequences, the arithmetic sequence holds a central position as a structured and fundamental pattern. As already discussed in chapter 1, an arithmetic sequence, often referred to as an arithmetic progression, is a sequence of numbers in which the difference between consecutive terms remains constant.

This constant difference is precisely defined as the "common difference" and is denoted by the symbol 'd.' Arithmetic sequences play a pivotal role in various mathematical contexts and find applications in fields such as finance, physics, and computer science, where linear and predictable changes are encountered (Chapman, Gotti, & Gotti, 2020).

Mathematically, an arithmetic sequence can be precisely defined through the following formula:

$$a_n = a_1 + (n-1)d$$

Here, the variables are as follows:

i. a_n represents the n th term of the sequence.

ii. a_1 denotes the first term of the sequence.

iii. n stands for the position of the term in the sequence.

iv. d signifies the common difference between consecutive terms.

The essence of an arithmetic sequence lies in its predictability. As one progresses from term to term, the common difference 'd' is consistently added to the previous term, ensuring uniformity throughout the sequence (Siagian, Aswin, & Herman, 2023).

Illustrating this with a mathematical example:

Example: Calculating the 20th Term

Suppose an arithmetic sequence is given with the first term a_1 equal to 3 and a common difference d of 4. To compute the 20th term, the arithmetic sequence formula is employed.

Using the formula:

$$a_{20} = a_1 + (20-1)\cdot d$$

Substituting the given values:

$$a_{20} = 3 + (20-1)\cdot 4$$

Simplifying the expression within the parentheses:

$$a_{20} = 3 + 19\cdot 4$$

Calculating the product:

$$a_{20} = 3 + 76$$

Finally, summing the values:

$$a_{20} = 79$$

Hence, the 20th term of this arithmetic sequence is determined to be 79.

Finding any specific term in an arithmetic sequence, such as the 20th term, is crucial for a variety of reasons. Firstly, it serves as a validation point, confirming that the sequence maintains its consistent pattern over the desired number of terms. This verification of pattern is not limited to the 20th term but can be applied to any term of interest. Additionally, by choosing a specific term, you gain insights into the sequence's behavior at that point in its progression. This understanding can be invaluable in real-world applications, such as financial planning or modeling, where you need to project future values based on the established pattern. Furthermore, selecting any term in the sequence requires the comprehensibility of mathematical concepts, making them more accessible and illustrative, especially in educational contexts (Rahmani-Andebili, 2021) (Figure 2.2).

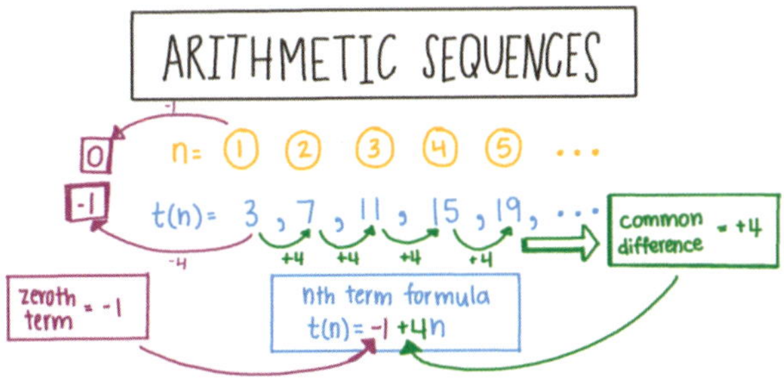

Figure 2.2. Example of an arithmetic sequence.

Source: Nagwa, Creative Commons License.

2.2.1. The Common Difference

The concept of the 'common difference' is significant in the context of arithmetic sequences. In the following sections, we will discuss this mathematical concept, thoroughly defining it, examining its implications, and discussing its intricacies through illustrative examples (Alexan et al., 2018).

Definition and Mathematical Expression: In the arithmetic sequence: $a_n = a_1 + (n-1)d$, each term differs from the previous one by a fixed constant. This constant is what we refer to as the 'common difference' denoted by (d) (Ross, 2017) (Figure 2.3).

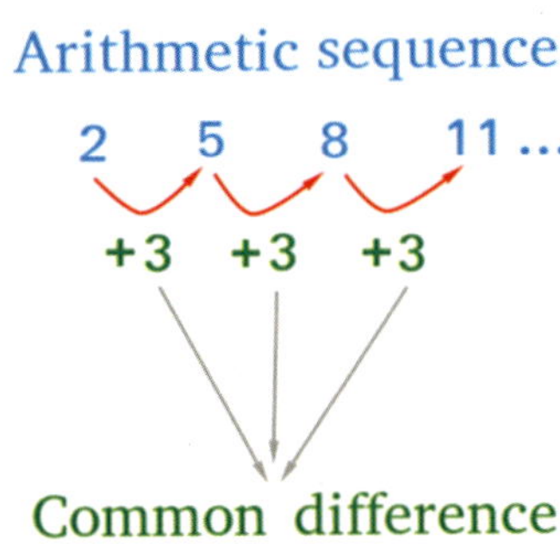

Figure 2.3. Schematic of common difference of arithmetic mean.

Source: Math-dictionary, Creative Commons License.

2.2.1.1. Visualizing the Linear Progression

Suppose a sequence of numbers where each term increases or decreases by the same constant number. This gradual shift creates a visually perceptible line when plotted on a graph. As each term is connected to the next through the common difference, the sequence unfolds linearly.

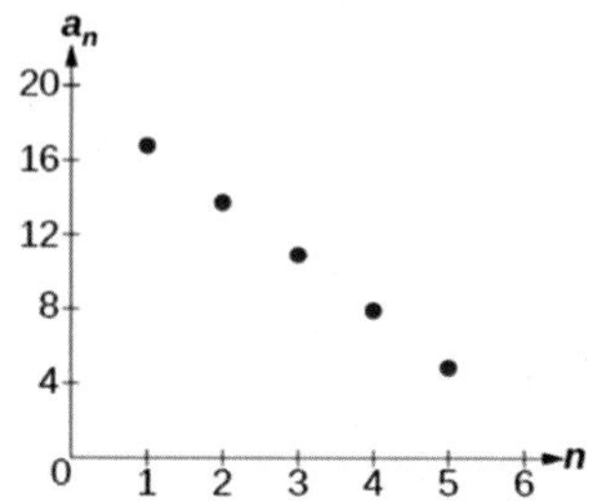

Figure 2.4. Graph illustrating tan arithmetic sequence with the common difference of -3.

Source: Lumen Learning, Creative Commons License.

The Figure 2.4 shows an arithmetic sequence with $a_1 = 17 \, and \, d = -3$.

Calculating Terms Using the Common Difference: Consider the arithmetic sequence $(5, 9, 13, 17, \ldots)$ with a common difference 4. We wish to determine the 10th term of this sequence using the common difference formula.

Using the formula of an arithmetic sequence,

$$a_{10} = 5 + (10 - 1) \cdot 4 = 5 + 36 = 41$$

Hence, the 10th term of this sequence is 41, showcasing the consistent progression governed by the common difference.

2.2.1.2. Applications in Real-World Scenarios

The common difference finds its application in various real-world contexts. Consider financial investments where the value increases or decreases by a fixed amount periodically. This incremental change aligns with the arithmetic sequence's linear growth pattern, wherein the constant difference represents the regular increment or decrement.

For instance, if an individual invests \$100 in a savings account with an annual interest rate of 5%, the balance at the end of each year forms an arithmetic sequence. The common difference in this example is $5, representing the annual interest earned.

The common difference concept enables us to address various problems involving arithmetic sequences. Whether it is predicting future terms, determining past terms, or understanding the relationship between different terms, the common difference acts as a guiding principle.

2.2.2. Finding the nth Term of an Arithmetic Sequence

Within the domain of arithmetic sequences, the determination of the nth term presents itself as a mathematical pursuit like the solving of a cryptic problem. This nth term holds the key to comprehending the ordered advancement of numbers within the sequence.

2.2.2.1. The Importance of the nth Term

The nth term of an arithmetic sequence serves as a pivotal tool for various mathematical and real-world applications. It allows us to predict or determine the value of any term in the sequence without the need to calculate all the preceding terms individually. This efficiency is especially valuable when dealing with lengthy sequences.

Illustrative Example: Consider the following arithmetic sequence: 7, 11, 15, 19, 23, … It is clear that the common difference, 'd,' between each term is 4. We want to find the 10th term of this sequence (a_{10}).

Using the nth term formula:

$$a_n = a_1 + (n-1)d$$

We can plug in the values for this sequence:

$a_1 = 7$ (the first term)

$n = 10$ (we want to find the 10th term)

$d = 4$ (the common difference)

Now, let's calculate (a_{10}):

$$a_{10} = 7 + (10-1) \cdot 4$$

$$a_{10} = 7 + 9 \cdot 4$$

$$a_{10} = 7 + 36$$

$$a_{10} = 43$$

So, the 10th term (a_{10}) of this arithmetic sequence is 43.

2.2.2.2. Insight into the Formula

To gain a deeper understanding of why this formula works, let's break it down step by step.

a_1 : This is the first term of the sequence and serves as the starting point.

$n-1$: This part accounts for how many "steps" we need to take within the sequence to reach the desired term. Since we're finding the nth term, we need to take $(n-1)$ steps because we start from the first term.

d : The common difference, (d), represents the magnitude of each step. Multiplying $(n-1)$ by (d) ensures we take the correct number of steps in the correct direction.

a_n : This is the nth term we are seeking, and it is obtained by adding the number of steps $(n-1)d$ to the starting point (a_1).

In essence, the formula accounts for the cumulative effect of taking a certain number of steps $(n-1)$ of a consistent size (d) from the starting point (a_1) to reach the nth term (a_n).

2.2.2.3. Applications in Real-World Scenarios

The ability to find the nth term of an arithmetic sequence is not confined to mathematical exercises alone.

It finds extensive applications in various real-world scenarios. Consider financial calculations, where periodic increments or decrements occur. For instance, if you have a savings account with an initial balance (a_1) and a fixed annual interest rate (d), you can use the nth term formula to predict the balance (a_n) after a certain number of years (n). This is a practical application of arithmetic sequences in finance (Firdaus et al., 2020).

Without the nth term formula, calculating a specific term in a long arithmetic sequence could be a laborious task, especially for sequences with thousands or even millions of terms. With this formula, you can swiftly determine any term of interest. The formula for finding the nth term of an arithmetic sequence is a powerful tool in mathematics and practical applications. It simplifies the process of predicting or determining specific terms within a sequence characterized by a constant common difference (Sarker, 2021).

2.3. ARITHMETIC MEAN AND SERIES

In mathematics, arithmetic means and arithmetic series are fundamental concepts that help us understand how numbers balance and add up in a sequence. This exploration aims to break down the arithmetic mean, dig into the arithmetic series, and show their importance through math formulas and real-world examples.

2.3.1. The Arithmetic Mean

The arithmetic mean, often simply referred to as the "average," is a fundamental concept in mathematics. It provides a central value that represents a set of numbers in such a way that the sum of deviations from this value is zero. In other words, it is the balance point of a data set. The arithmetic mean of a set of numbers is calculated by adding up all the numbers and then dividing by the count of numbers in the set (Casey et al., 2004). Mathematically, the arithmetic mean $(\overline{x})$ of a set of (n) numbers $(x_1, x_2, x_3, \ldots, x_n)$ is expressed as:

$$\overline{x} = \frac{1}{n}(x_1 + x_2 + x_3 + \ldots + x_n)$$

This formula finds its practical utility in various fields, such as statistics, finance, and science, where understanding the central tendency of data is paramount.

2.3.2. The Arithmetic Series

An arithmetic series, on the other hand, is the summation of the terms of an arithmetic sequence. It provides a concise representation of the cumulative effect of an ordered list of numbers that progress uniformly with a common difference (Alghar et al., 2022; Anjum & Mukherjee, 2022).

The sum of the first (n) terms of an arithmetic series, denoted as (S_n), can be expressed as follows:

$$S_n = n \cdot \overline{x}$$

Where:

S_n represents the sum of the first (n) terms.

n is the number of terms.

$\overline{x}$ is the arithmetic mean of the series.

This formula offers a straightforward means of calculating the total value of an arithmetic series, making it a valuable tool in various mathematical and real-world contexts.

- **Example: Calculating the Arithmetic Mean**

Consider a set of exam scores: 85, 90, 78, 92, and 88. We want to find the arithmetic mean of these scores.

Using the arithmetic mean formula:

$$\overline{x} = \frac{1}{n}(x_1 + x_2 + x_3 + x_4 + x_5)$$

Where (n) is the count of numbers in the set (in this case $n=5$, and x_1, x_2, x_3, x_4, x_5 are the individual scores, we can calculate the mean as follows:

$$\overline{x} = \frac{1}{5}(85 + 90 + 78 + 92 + 88)$$

Now, add up the numbers:

$$\overline{x} = \frac{1}{5}(433)$$

Finally, calculate the mean:

$$\overline{x} = \frac{433}{5} = 86.6$$

So, the arithmetic mean of these exam scores is 86.6.

- **Example: Calculating the Arithmetic Series**

Suppose the car's speedometer readings at regular intervals are as follows: 60 km/h, 65 km/h, 70 km/h, and 75 km/h. We want to find the total distance covered in 4 hours.

First, let's calculate the arithmetic mean of the speeds using the formula we discussed earlier. In this case, $(n=4)$, and the speeds are 60, 65, 70, and 75 km/h:

$$\bar{x} = \frac{1}{4}(60+65+70+75) = \frac{1}{4}(270) = 67.5 \text{ km/h}$$

Now that we have the arithmetic mean of 67.5 km/h, we can calculate the total distance (S_n) covered in 4 hours using the arithmetic series formula:

$$S_n = n \cdot \bar{x} = 4 \cdot 67.5 = 270 \text{ km}$$

So, the car covered a total distance of 270 kilometers in 4 hours.

2.3.3. Applications of Arithmetic Mean and Series

The concepts of arithmetic mean and arithmetic series find application in various fields (Pickover, 2011) (Figure 2.5):

1. **Statistics:** In data analysis, the arithmetic mean is a fundamental measure of central tendency, providing insights into the average value of a dataset.
2. **Finance:** In finance, the arithmetic means is used to calculate returns on investments, helping investors understand the average return on their portfolio.
3. **Physics:** In physics, the arithmetic mean is employed to calculate average velocities and accelerations (Joan, 2015).
4. **Engineering:** In engineering, the arithmetic mean is used to determine the average stress or strain in materials.
5. **Economics:** In economics, the arithmetic mean is used to calculate indices like the Consumer Price Index (CPI), reflecting the average change in prices over time.
6. **Distance and Speed Calculations:** As demonstrated in the car trip example, the arithmetic mean and series help calculate total distances and cumulative effects of varying speeds (Grassi, 2021; Sun & Chen, 2018).

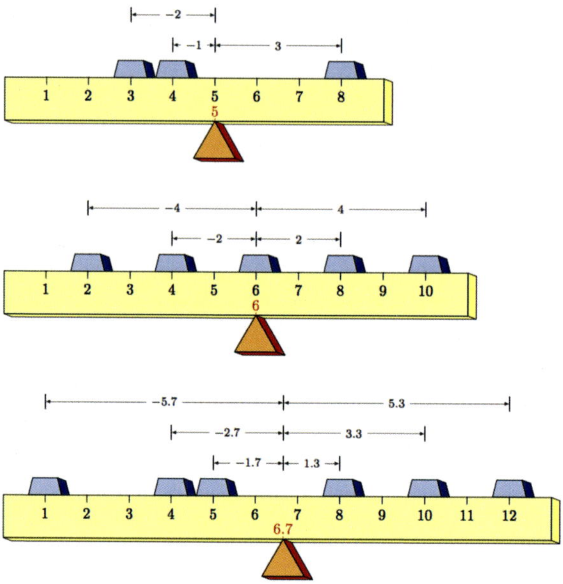

Figure 2.5. The arithmetic mean can be visualized as a balancing point on a scale.

Source: Brillinat.org, Creative Commons License.

2.3.4. Arithmetic Mean and Series in Finance

Arithmetic sequences find invaluable applications in finance, where they help model and solve various financial scenarios. These sequences provide a structured approach to understanding how values change over time, particularly when there is a consistent change or growth in financial parameters (Wang et al., 2012).

Remember

In finance, the arithmetic series plays a crucial role in calculating the total amount paid overtime for loans or investments with fixed, regular payments. It's a fundamental concept for financial planning and understanding the cost or return on investments.

- **Arithmetic Mean:** It provides a central measure of a dataset and is used in finance to understand the typical or average value of financial data. For example, consider the monthly returns of a stock over a year: +2%, -1%, +4%, -3%, +5%, and so on. To find the average monthly return, you add up all these returns and divide them by the number of months (12, in this case). So

$,(2-1+4-3+5+.)/12$ = Average Monthly Return.

- **Arithmetic Series**: In finance, an arithmetic series is a summation of financial data that follows a consistent linear pattern. It is used to calculate the total amount or accumulated value of a series of cash flows over time, where each cash flow differs by a constant amount.

Suppose you are running a small business, and every month, you make a profit that increases by \$500. In the first month, your profit is \$1,000, and it increases by \$500 each month. You want to calculate your total profit for the first six months using the arithmetic series formula.

Using the formula $(S_n = n \cdot \overline{x})$,

$n=6$ (as we are calculating the first six months)

and

$$\left(\overline{x} = \frac{\text{First-month profit} + \text{Last month's profit}}{2} = \frac{1,000+3,000}{2} = 2,000\right)$$

Now, plug these values into the formula:

$$S_n = 6 \cdot 2,000 = 12,000$$

So, your total profit for the first six months would be \$12,000.

Furthermore, arithmetic series in finance can be extended to calculate the total interest earned over a specific period. The total interest is the sum of the annual interest earned each year, which is a simple arithmetic series. For instance, if you want to know how much interest you will earn after 10 years, you can sum the first 10 terms of the arithmetic sequence, each representing the interest earned in a specific year (Gunnarsson et al., 2016). Arithmetic sequences in finance offer a systematic way to model the growth or change of financial values over time. Whether it is calculating future investment values, projecting savings growth, or determining interest earnings, these sequences provide financial professionals and individuals with powerful tools for making informed financial decisions and achieving their financial goals (Aharonov & Stephenson, 1998).

2.4. GEOMETRIC SEQUENCES

These sequences exhibit an inherent multiplicative pattern, showcasing exponential growth or decay. In this section, geometric sequences will be discussed, showing their fundamental characteristics, mathematical expressions, and practical applications (Figure 2.6).

A geometric sequence is an ordered list of numbers where each term is obtained by multiplying the previous term by a fixed constant known as the common ratio, often denoted as (r). The common ratio dictates the progression of the sequence, making it either increase exponentially when (r) is greater than 1 or decrease exponentially when (r) is between 0 and 1 (Morton et al., 2017).

Mathematically, a geometric sequence can be defined as:

$$a_n = a_1 \cdot r^{(n-1)}$$

Here, a_n represents the nth term of the sequence, a_1 is the first term, r is the common ratio, and n signifies the term number.

- **Example: Geometric Growth of Bacteria**

To illustrate the concept of geometric sequences, consider a situation where a bacterium population doubles every hour. Starting with an initial population of 10 bacteria $(a_1 = 10)$, we can use a geometric sequence to model its growth (Gunnarsson et al., 2016).

The common ratio in this scenario is $(r=2)$ because the population doubles each hour. Using the formula for a geometric sequence, we can calculate the population at different time intervals:

1. After 1 hour $(n=2):(a_2 = 10 \cdot 2^{(2-1)} = 20)$ bacteria.
2. After 2 hours $(n=3):(a_3 = 10 \cdot 2^{(3-1)} = 40)$ bacteria.
3. After 3 hours $(n=4):(a_4 = 10 \cdot 2^{(4-1)} = 80)$ bacteria.

This progression demonstrates the exponential growth characteristic of geometric sequences when the common ratio is greater than 1 (Amsari et al., 2022).

- **Example: Geometric Decay**

Now, consider a scenario where the value of a car depreciates by 20% each year. Starting with a car's initial value of $20,000 $(a_1 = 20,000)$, we can use a geometric sequence to model its depreciation.

The common ratio in this case is $(r=0.8)$ because the car's value decreases to 80% of its previous value each year. Using the formula for a geometric sequence, we can calculate the car's value after different years:

1. After 1 year $(n=2):\left(a_2 = 20,000 \cdot 0.8^{(2-1)} = 16,000\right)$ dollars.
2. After 2 years $(n=3):\left(a_3 = 20,000 \cdot 0.8^{(3-1)} = 12,800\right)$ dollars.
3. After 3 years $(n=4):\left(a_4 = 20,000 \cdot 0.8^{(4-1)} = 10,240\right)$ dollars.

This demonstrates the exponential decay behavior of geometric sequences when the common ratio is between 0 and 1.

- **Applications in Real-world Scenarios:** Geometric sequences find extensive application in various real-world contexts due to their ability to model exponential growth and decay. Here are a few examples:

1. **Finance:** In investments, geometric sequences help calculate compound interest, where the initial amount grows by a fixed percentage over time.
2. **Biology:** Geometric sequences are used to model population growth or decline in ecological studies, such as the spread of viruses or the decline of endangered species.
3. **Physics:** They play a role in modeling the exponential decay of radioactive substances, as well as in understanding processes like sound attenuation and light intensity.
4. **Engineering:** Engineers use geometric sequences to analyze signal processing, filter design, and the behavior of electrical circuits.
5. **Computer Science:** In algorithms and data structures, geometric sequences can represent the time complexity of algorithms with exponential growth or decay.

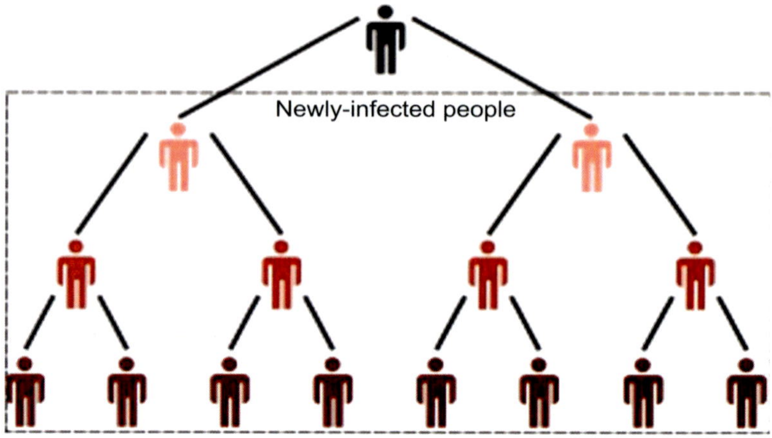

Figure 2.6. An example of the geometric sequence where each person infects two more people with the flu virus.

Source: Siyavula, Creative Commons License.

- **The Power of Geometric Sequences:** Geometric sequences hold a unique place in the world of mathematical patterns. Their exponential nature allows them to accurately represent phenomena involving constant growth or decay. Understanding geometric sequences equips us with a powerful tool for predicting and modeling real-world scenarios, from investments to population dynamics, where the power of exponential change is at play (Shih & Lee, 2019).

2.4.1. The Common Ratio

Common ratio plays a pivotal role, guiding the sequence's behavior and dictating whether it showcases exponential growth or decay. In this section, we will discuss the concept of the common ratio and its mathematical implications and provide concrete examples to show its significance (Kondo & Miura, 2010).

2.4.1.1. Defining the Common Ratio

The common ratio (r) in a geometric sequence is the fixed constant by which each term is obtained from the preceding one. In simpler terms, it represents the factor by which the sequence either grows or shrinks from term to term. This constant is integral to the geometric sequence's identity, as it fundamentally defines its progression (Figure 2.7).

Mathematically, for common ratio, we can use geometric sequence:

$$a_n = a_1 \cdot r^{(n-1)}$$

Here, (a_n) denotes the (n)-th term of the sequence, (a_1) signifies the first term, $(\boldsymbol{r})$ **is the common ratio**, and (n) represents the term number.

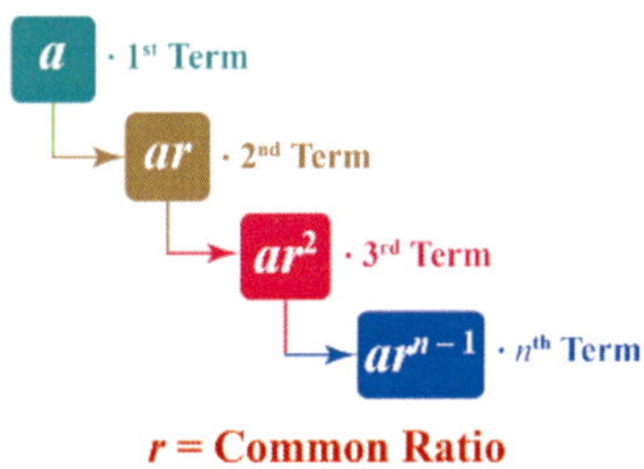

Figure 2.7. Illustration of common ratio.

Source: Kofa study, Creative Commons License.

2.4.1.2. The Common Ratio's Impact on the Sequence

The magnitude and sign of the common ratio (r) significantly influence the nature of the geometric sequence:

1. **Exponential Growth** $(\boldsymbol{r}>1)$**:** When the common ratio (r) is greater than 1, the sequence exhibits exponential growth.

   Each term is larger than the previous one by a factor of (r). This behavior is characteristic of phenomena involving rapid expansion, such as population growth or compound interest.

2. **Exponential Decay** $(0< \boldsymbol{r}<1)$**:** Conversely, when the common ratio (r) falls between 0 and 1, the sequence undergoes exponential decay. Each term becomes smaller than the preceding one by a factor of (r). This pattern is observed in processes like radioactive decay or diminishing returns in investments.

3. **No Growth or Decay $(r=1)$:** In the special case where the common ratio (r) equals 1, the sequence remains constant. In this scenario, each term is equal to the preceding term, signifying no growth or decay.

- **Example: Exponential Growth**

Consider a scenario where you invest $1,000 in a savings account that provides an annual interest rate of 10%. Using the formula for a geometric sequence, we can model the growth of your investment over several years (Johansen & Sornette, 2001).

In this example, (a_1) represents the initial investment of $1,000, and the common ratio (r) is 1 plus the annual interest rate (expressed as a decimal), which is $(0.10+1=1.10)$. We want to find the value of the investment after different years.

After 1 year $(n=2)$, the investment becomes:

$a_2 = 1,000 \cdot 1.10^{(2-1)} = 1,100$ dollars.

After 2 years $(n=3)$, it grows to:

$a_3 = 1,000 \cdot 1.10^{(3-1)} = 1,210$ dollars.

After 3 years $(n=4)$, it further increases to:

$a_4 = 1,000 \cdot 1.10^{(4-1)} = 1,331$ dollars.

This sequence demonstrates the concept of exponential growth, where each term is calculated by multiplying the previous term by the common ratio of 1.10.

- **Example: Exponential Decay**

Suppose you have a radioactive substance with an initial mass of 1 gram, and it decays at a constant rate of 20% per hour. We can model the decay of this substance over time (Arroyo et al., 2011).

In this scenario, (a_1) represents the initial mass of 1 gram, and the common ratio (r) is 1 minus the decay rate (expressed as a decimal), which is (0.20). We aim to find the remaining mass after different hours.

After 1 hour $(n=2)$, the mass becomes:

$a_2 = 1 \cdot (1-0.20)^{(2-1)} = 0.80$ grams.

After 2 hours $(n = 3)$, it further decreases to:

$a_3 = 1 \cdot (1 - 0.20)^{(3-1)} = 0.64$ grams.

After 3 hours $(n = 4)$, it continues to decay:

$a_4 = 1 \cdot (1 - 0.20)^{(4-1)} = 0.512$ grams.

This sequence illustrates the concept of exponential decay, where each term is calculated by multiplying the previous term by the common ratio of 0.80.

The common ratio is a fundamental parameter that underpins the behavior of these sequences, enabling us to model and understand phenomena involving consistent change over time. From finance to biology, physics to engineering, and beyond, the concept of the common ratio in geometric sequences stands as a powerful tool for describing the dynamics of exponential processes in the natural and mathematical world (Barndorff-Nielsen & Shephard, 2001).

2.4.1.3. Determining the 10th Term

To further illustrate the concept of the common ratio within geometric sequences, let's explore a specific example to determine the 10th term of a geometric sequence. This example will highlight how the common ratio governs the progression of the sequence (Gura, 1996).

- **Scenario:** Consider a geometric sequence where the first term (a_1) is 3, and the common ratio (r) is 2. We want to find the 10th term (a_{10}) of this sequence.

Using the formula for a geometric sequence:

$$a_n = a_1 \cdot r^{(n-1)}$$

Here, we have:

$a_1 = 3$ (the first term)

$r = 2$ (the common ratio)

$n = 10$ (we want to find the 10th term)

Now, let's calculate (a_{10}) using the formula:

$$a_{10} = 3 \cdot 2^{(10-1)}$$

Simplify the exponent:

$$a_{10} = 3 \cdot 2^9$$

Calculate $\left(2^9\right)$, which is 512:

$$a_{10} = 3 \cdot 512$$

Finally, multiply 3 by 512 to find the 10th term:

$$a_{10} = 1{,}536$$

So, the 10th term $\left(a_{10}\right)$ of this geometric sequence with a first term of 3 and a common ratio of 2 is equal to 1,536.

In this example, the common ratio $\left(r = 2\right)$ played a crucial role in determining the 10th term. It represented the factor by which each term was obtained from the previous one. As we can see, the progression was not linear; each term was twice the value of the preceding term, leading to exponential growth. This is a characteristic of geometric sequences when the common ratio is greater than 1 (Qonita & Wuryandari, 2015). The 10th term, 1,536, is a testament to the power of exponential growth. Starting with the first term of 3, each subsequent term grew significantly larger due to the common ratio of 2. This property is why geometric sequences are valuable for modeling processes involving exponential change, such as investments with compound interest or population growth (Qonita et al., 2015).

2.4.2. Calculating the nth Term of a Geometric Sequence

This capability empowers us to determine the value of any term in the sequence without having to compute all the preceding terms individually. In this exploration, we will discuss the methodology for calculating the nth term of a geometric sequence, presenting the mathematical framework behind it and providing illustrative examples.

Tip

When calculating the nth term of a geometric sequence $\left(a\, sub\, n = a\, sub 1 * r^{(n-1)}\right)$, pay close attention to the common ratio (r) and the initial term (a sub 1). Small changes in these values can lead to significant differences in the sequence's behavior.

2.4.2.1. The General Formula

To calculate the nth term of a geometric sequence, we utilize the general formula provided above. This formula encapsulates the essence of geometric sequences and allows us to compute any term within the sequence with ease (Angraini et al., 2023).

- Calculating the nth Term:

Consider the following geometric sequence: 2, 6, 18, 54, 162, …

In this sequence:

i. The first term (a_1) is 2.

ii. The common ratio (r) is found by dividing any term by the preceding term. For instance, $\left(\frac{6}{2}=3\right), \left(\frac{18}{6}=3\right)$, and so on. Therefore, $(r=3)$.

Now, suppose we want to find the 7th term (a_7) of this sequence. We can use the general formula:

$$a_n = a_1 \cdot r^{(n-1)}$$

In this case:

$$a_1 = 2$$

$$r=3$$

$$n=7$$

Now, let's calculate (a_7):

$$a_7 = 2 \cdot 3^{(7-1)}$$

Simplify the exponent:

$$a_7 = 2 \cdot 3^6$$

Calculate (3^6), which is 729:

$$a_7 = 2 \cdot 729$$

Finally, multiply 2 by 729 to find the 7th term:

$$a_7 = 1458$$

So, the 7th term (a_7) of this geometric sequence is 1458.

2.4.2.2. Understanding the Formula

The formula for calculating the nth term of a geometric sequence is derived from the inherent multiplicative pattern within these sequences. Each term is a result of multiplying the preceding term by the common ratio (r). Therefore, to find any term in the sequence, we start with the first term (a_1), then apply the common ratio (r) repeatedly for $(n-1)$ times, as the term number dictates (Tong et al., 2021). Calculating the nth term of a geometric sequence is fundamental and enables one to understand and model exponential growth or decay patterns in various domains.

The formula, derived from the inherent multiplicative nature of these sequences, simplifies the process of determining specific terms within the sequence, making it a valuable tool for solving practical problems and making informed decisions. Geometric sequences, with their predictable progression, continue to be a cornerstone in mathematics and its applications, showcasing the elegance and utility of mathematical patterns (Aharonov & Stephenson, 1998).

2.4.3. Geometric Mean and Series

These mathematical concepts provide us with profound insights about multiplicative relationships, equilibrium, and the summation of geometric progressions. In this section, we will study geometric mean, explore the geometric series, and present their significance through mathematical formulations, real-world examples, and practical applications (Na, Sun, Ingraham, & Yoon, 2017).

2.4.3.1. Geometric Mean

Geometric mean is a central concept in mathematics that captures the essence of multiplicative relationships. Unlike the arithmetic mean, which represents the central value in a set of numbers, the geometric mean represents the central tendency when numbers are related by multiplication rather than addition (Gong, 2019).

Mathematically, the geometric mean (G) of (n) positive numbers $(x_1, x_2, x_3, \ldots, x_n)$ is defined as:

$$G = \sqrt[n]{x_1 \cdot x_2 \cdot x_3 \cdot \ldots \cdot x_n}$$

Here, (n) denotes the number of values, and $(\sqrt[n]{\ldots})$ represents the nth root, which is equivalent to raising the product of the numbers to the power of $(1/n)$. The geometric mean provides an essential tool for finding the average of numbers that are multiplied together or are part of a geometric progression (Arbib & Manes, 1982).

- **Calculating the Geometric Mean:**

Consider a scenario where you are measuring the growth rate of a plant over several days. The plant's height on each day is as follows (in centimeters): 2, 4, 8, 16, 32, and 64. We want to find the geometric mean of these measurements.

Using the formula for the geometric mean, we can calculate it as follows:

$$G = \sqrt[6]{2 \cdot 4 \cdot 8 \cdot 16 \cdot 32 \cdot 64}$$

Simplify the product inside the nth root:

$$G = \sqrt[6]{262{,}144}$$

Now, calculate the sixth root of 262,144:

$$G = \sqrt[6]{262{,}144} \approx 32$$

So, the geometric mean of the plant's height measurements is approximately 32 centimeters.

2.4.3.2. The Geometric Series

A geometric series is the summation of the terms of a geometric sequence. It provides a concise representation of the cumulative effect of multiplying each term in a sequence by a fixed constant, known as the common ratio (r). The sum of the first (n) terms of a geometric series, denoted as (S_n), can be expressed as follows:

$$S_n = \frac{a_1\left(1 - r^n\right)}{1 - r}$$

where: S_n represents the sum of the first (n) terms; a_1 is the first term of the sequence; r is the common ratio; and n is the number of terms.

This formula provides a systematic way to calculate the total value of a geometric series, making it a valuable tool in various mathematical and real-world contexts.

- **Calculating the Geometric Series:**

To further illustrate the geometric series, let's consider a scenario where we want to calculate the sum of the first 5 terms of a geometric series. The first term (a_1) is 3, and the common ratio (r) is 2 (Yu et al., 2017).

Using the formula for the sum of a geometric series, we can calculate it as follows:

$$S_5 = \frac{3(1-2^5)}{1-2}$$

Simplify the numerator:

$$S_5 = \frac{3(1-32)}{-1}$$

Further simplify:

$$S_5 = \frac{3(-31)}{-1}$$

Now, calculate the final result:

$$S_5 = \frac{-93}{-1} = 93$$

So, the sum of the first 5 terms of this geometric series with a first term of 3 and a common ratio of 2 is 93.

The geometric series formula encapsulates the cumulative effect of adding terms from a geometric sequence. It accounts for the progression of the sequence as each term is obtained by multiplying the preceding term by the common ratio (r). The formula provides a convenient way to find the total value of the series without having to compute each term individually (DiDomenico, 1997). The denominator in the formula $(1-r)$ ensures that the sum converges to a finite value when $(|r|<1)$. In cases where $(|r|\geq 1)$, the series diverges, meaning it does not have a finite sum. Hence, the geometric mean and geometric series, with their focus on multiplicative relationships and cumulative effects, are foundational concepts in mathematics and its applications. The geometric mean provides a means to find the average in situations involving multiplication, while the geometric series simplifies the summation of geometric progressions (Shute et al., 2006).

2.4.3.3. Modeling Population Growth

Population growth is a complex and vital field of study in various disciplines, including biology, ecology, and demographics. Geometric mean and series play a significant role in modeling population growth, as they capture the multiplicative relationships inherent in this process. In this example, we will explore how geometric mean and series are used to model population growth, providing both mathematical foundations and practical insights.

When a population grows under certain conditions, the number of individuals in each generation is typically related to the number in the previous generation through a fixed factor known as the reproductive rate or growth rate (Gutierrez et al., 2018).

Let's denote the population in each generation as (P_n), where (n) is the generation number. The relationship between (P_n) and (P_{n-1}) can be expressed as:

$$P_n = r \cdot P_{n-1}$$

Here, (r) is the reproductive rate, representing the factor by which the population grows from one generation to the next.

- **Calculating the Reproductive Rate:**

The geometric mean plays a crucial role in determining the reproductive rate (r). To calculate (r), we use the geometric mean of the ratios of population sizes between consecutive generations.

Suppose we have observed the population sizes for several consecutive generations, say $(P_1, P_2, P_3, \dots, P_n)$. The reproductive rate (r) can be calculated as follows:

$$r = \sqrt[n-1]{\frac{P_2}{P_1} \cdot \frac{P_3}{P_2} \cdot \frac{P_4}{P_3} \cdot \ldots \cdot \frac{P_n}{P_{n-1}}}$$

Here, we calculate the geometric mean of the ratios of population sizes from one generation to the next. This gives us the average growth factor, which describes the multiplicative relationship between generations (Ulam, 1962).

- **Example:**

Suppose we have data on a population of rabbits over five generations:

$P_1 = 100$ (initial population)

$P_2 = 200$ (after the first generation)

$P_3 = 4($ (after the second generation)

$P_4 = 800$ (after the third generation)

$P_5 = 1600$ (after the fourth generation)

To find the reproductive rate (r), we can apply the formula for the geometric mean:

$$r = \sqrt[4]{\frac{200}{100} \cdot \frac{400}{200} \cdot \frac{800}{400} \cdot \frac{1600}{800}}$$

Simplify the ratios:

$$r = \sqrt[4]{2 \cdot 2 \cdot 2 \cdot 2}$$

Calculate the fourth root:

$$r = \sqrt[4]{16} = 2$$

So, in this example, the reproductive rate (r) for the rabbit population is 2, indicating that each generation's population is twice the size of the previous generation.

Once we have determined the reproductive rate (r), we can use it to model population growth over time using a geometric series. The formula for calculating the population size (P_n) at any given generation (n) is:

$$P_n = P_1 \cdot r^{(n-1)}$$

Here, (P_1) is the initial population size, (r) is the reproductive rate, and (n) is the generation number.

This formula allows us to project the population size for any future generation, given the initial population size and the reproductive rate (Grassi, 2021). Let's continue with our rabbit population example. We already calculated the reproductive rate $(r = 2)$.

Suppose we want to predict the rabbit population size after 10 generations, starting with an initial population of 100.

Using the formula for population growth:

$$P_n = 100 \cdot 2^{(n-1)}$$

Substitute $(n=10)$ and calculate:

$$P_{10} = 100 \cdot 2^{(10-1)}$$

Simplify the exponent:

$$P_{10} = 100 \cdot 2^9$$

Calculate $\left(2^9\right)$, which is 512:

$$P_{10} = 100 \cdot 512$$

Finally, multiply 100 by 512 to find the predicted population after 10 generations:

$$P_{10} = 51,200$$

So, according to our model, the rabbit population is expected to reach 51,200 individuals after 10 generations, starting with an initial population of 100 and a reproductive rate of 2 (Genkin et al., 2017).

Hence, the geometric mean and series are indispensable tools for modeling population growth, a fundamental aspect of various scientific and societal problems. By understanding the multiplicative relationships between generations and utilizing the geometric mean to calculate the reproductive rate, we can project and predict population sizes over time. These mathematical concepts bridge theory and practice, facilitating our comprehension of population dynamics and aiding in decision-making processes across multiple disciplines.

2.5. COMBINING ARITHMETIC AND GEOMETRIC SEQUENCES

Combining arithmetic and geometric sequences is a powerful mathematical approach. In such sequences, each term's difference forms an arithmetic sequence, while the ratios between terms create a geometric sequence. This synergy is exemplified in financial calculations like annuities, where regular deposits (arithmetic) accumulate with compounded interest (geometric). For instance, in a monthly savings plan with \$100 deposits (arithmetic) and an annual interest rate of 5% (geometric), the total savings over time expresses the appropriate blend of both sequences, yielding a comprehensive understanding of financial growth. This fusion of mathematical patterns finds applications in various fields, facilitating precise modeling and analysis (Sari & Susanti, 2020).

2.5.1. Transitioning Between Sequences

The seamless transition between arithmetic and geometric sequences represents a mathematical sequence that exploits the interplay between these two distinct yet interlinked sequences. This transition often involves identifying the point where an arithmetic sequence transforms into a geometric one or vice versa, leading to interesting patterns and insights. In this section, we will discuss the intricacies of transitioning between sequences, showing mathematical formulations and illustrative examples to express this captivating phenomenon (Vaidyanathan, 2014).

2.5.1.1. Recognizing the Transition Point

The transition between arithmetic and geometric sequences occurs when the common difference of the arithmetic sequence transforms into the common ratio of the geometric sequence or vice versa. This turning point marks where the linear progression of the arithmetic sequence changes into the exponential growth or decay of the geometric sequence and vice versa (Zazkis & Liljedahl, 2002).

Mathematically, the transition point in an arithmetic sequence with a common difference (d) and a geometric sequence with a common ratio (r) can be found by equating the respective formulas for the nth terms:

$$a_n^{\text{arithmetic}} = a_1^{\text{arithmetic}} + (n-1) \cdot d$$

$$a_n^{\text{geometric}} = a_1^{\text{geometric}} \cdot r^{(n-1)}$$

Equating these expressions leads to:

$$a_1^{\text{arithmetic}} + (n-1)\cdot d = a_1^{\text{geometric}} \cdot r^{(n-1)}$$

Solving for (n) at this point signifies the transition between sequences.

- **Example: Identifying the Transition Point**

Consider the arithmetic sequence: 3, 7, 11, 15, … with a common difference of 4. Simultaneously, imagine the geometric sequence: 2, 6, 18, 54, … with a common ratio of 3. Let's determine the point at which these two sequences transition (Dwyer et al., 1998).

Using the formula for the nth term of an arithmetic sequence:

$$a_n^{\text{arithmetic}} = 3 + (n-1)\cdot 4$$

And for the n th term of a geometric sequence:

$$a_n^{\text{geometric}} = 2 \cdot 3^{(n-1)}$$

Setting these expressions equal

$$3 + (n-1)\cdot 4 = 2 \cdot 3^{(n-1)}$$

Solve for (n) by simplifying the equation:

$$3n - 3 = 2 \cdot 3^{(n-1)}$$

$$3n = 2 \cdot 3^{(n-1)} + 3$$

$$n = \frac{2 \cdot 3^{(n-1)} + 3}{3}$$

Calculating (n) yields the transition point where the two sequences merge.

Transitioning between sequences offers a unique perspective on the continuity of mathematical patterns. The shift from an arithmetic sequence's linear increase to a geometric sequence's exponential growth (or vice versa) illustrates how numbers evolve, leading to diverse applications in various fields (Kondo & Miura, 2010).

2.5.1.2. Applications in Financial Mathematics

The transition point is particularly relevant in financial mathematics. Consider investments that begin with regular deposits (arithmetic) and gradually accumulate compounded interest (geometric). Identifying the transition point allows precise calculations of when the investment structure changes and the impact on returns. Such insights enhance investment strategy and decision-making (Duan & Babu, 2007).

Transitioning between arithmetic and geometric sequences signifies the link where linear progression transforms into exponential growth or decay and vice versa. This mathematical phenomenon holds relevance in diverse applications, including finance, where it aids in dissecting financial structures with changing dynamics. Understanding this transition enriches our comprehension of sequences' intricacies and extends the utility of these concepts across various disciplines.

2.5.2. Problem-Solving with Mixed Sequences

Combining both arithmetic and geometric elements presents intriguing problem-solving opportunities across various mathematical and practical scenarios.

These sequences encapsulate the synergy of linear progression (arithmetic) and exponential growth or decay (geometric), making them valuable tools for tackling complex real-world challenges (Rachma et al., 2022).

2.5.2.1. Defining Mixed Sequences

Mixed sequences, as the name suggests, encompass a blend of arithmetic and geometric elements within a single sequence. They exhibit a pattern where each term is influenced by both an arithmetic common difference and a geometric common ratio. Such sequences often arise when a problem incorporates both linear and exponential factors (Amsari et al., 2022; Rachma et al., 2022).

Mathematically, a mixed sequence can be expressed as:

$$a_n = a_1 + (n-1)\cdot d + a_1 \cdot r^{(n-1)}$$

Here, (a_n) represents the nth term of the mixed sequence, (a_1) is the first term, (d) is the arithmetic common difference, (r) is the geometric common ratio, and (n) is the term number (Kendall & Hill, 1953).

2.5.2.2. Problem-Solving Approach

Solving problems involving mixed sequences requires a structured approach that leverages the distinctive characteristics of arithmetic and geometric components. The key steps are as follows:

1. **Identify the Elements:** Begin by identifying the elements of the mixed sequence:

   a_1 : The first term.

   d : The arithmetic common difference.

   r : The geometric common ratio.

   n : The term number of interest.

2. **Formulate the Mixed Sequence:** Utilize the mixed sequence formula to represent the problem context mathematically. This involves combining the arithmetic and geometric elements.
3. **Analyze the Problem:** Understand the problem statement and what it seeks to find or describe. Pay attention to whether the sequence exhibits growth (positive r) or decay $(0 < r < 1)$.
4. **Apply the Sequence Formula:** Plug in the known values into the mixed sequence formula and perform calculations to solve for the desired term (a_n) or other relevant quantities.

- **Example: Compound Interest**

Consider a savings account with an initial deposit of $1,000 (the first term, a_1 that accrues annual interest. The account has an annual interest rate of 5% (the geometric common ratio, (r) and an additional $100 (the arithmetic common difference, (d) deposited at the beginning of each year. We want to determine the balance after 5 years (Bolker & Pacala, 1997).

Using the mixed sequence formula:

$$a_n = a_1 + (n-1) \cdot d + a_1 \cdot r^{(n-1)}$$

Where:

$$a_1 = 1000$$

$d = 100$

$r = 1.05$ (5% annual interest rate)

$n = 5$ (after 5 years)

Substitute these values into the formula:

$$a_5 = 1000 + (5-1) \cdot 100 + 1000 \cdot (1.05)^{(5-1)}$$

Simplify:

$$a_5 = 1000 + 4 \cdot 100 + 1000 \cdot 1.05^4$$

Calculate $\left(1.05^4\right)$ as approximately 1.21550625:

$$a_5 = 1000 + 400 + 1000 \cdot 1.21550625$$

Perform the final calculation:

$$a_5 \approx 1000 + 400 + 1215.50625 \approx 2615.51$$

So, the balance in the savings account after 5 years will be approximately \$2,615.51.

2.5.2.3. Applications of Mixed Sequence

Mixed sequences find extensive applications in finance, growth models, and other fields. Some examples include:

1. **Investment Analysis:** Mixed sequences are employed to model investments with regular deposits and varying interest rates, facilitating precise financial planning.
2. **Population Growth:** In demographic studies, mixed sequences can represent population dynamics, where both linear factors (e.g., births) and exponential factors (e.g., exponential growth) influence the population's evolution (Dueñas-Osorio, 2005).
3. **Epidemiology:** Epidemiologists use mixed sequences to model the spread of diseases, accounting for both linear factors (e.g., new infections) and exponential factors (e.g., transmission rates).
4. **Project Management:** In project scheduling, mixed sequences can represent tasks with both fixed durations (arithmetic) and dependencies (geometric), aiding in project planning and resource allocation.

2.5.2.4. A Problem Involving Both Sequences

Let's explore a practical problem that involves both arithmetic and geometric sequences, highlighting the problem-solving capabilities of mixed sequences (Kanerva, 2009).

- **Problem Statement:** Suppose you have a savings goal of \$10,000, and you plan to achieve it by making regular monthly deposits into a savings account. The account offers an annual interest rate of 4%, compounded monthly. You decide to increase your monthly deposit by \$50 every month. How many months will it take for your savings to reach or exceed \$10,000?
- **Problem-Solving Approach:** This problem seamlessly combines arithmetic and geometric sequences. The increasing monthly deposit represents an arithmetic sequence, while the compounding interest on the growing savings embodies a geometric sequence. Let's solve it step by step:

1. **Identify the Elements:**

i. Initial savings (a_1): \$0 (since there is no initial savings).

ii. Monthly deposit increment (d): \$50.

iii. Monthly interest rate (r): $\left(\frac{4\%}{12} = 0.3333\%\right)$ or (0.003333) (expressed as a decimal).

iv. Savings goal $((S))$: \$10,000.

2. **Formulate the Mixed Sequence:**

We can use the mixed sequence formula to represent the savings over time (a_n):

$$a_n = a_1 + (n-1) \cdot d + a_1 \cdot r^{(n-1)}$$

3. **Analyze the Problem:**

We want to find the number of months it takes for the savings (a_n) to reach or exceed \$10,000.

4. **Apply the Sequence Formula:**

Substitute the known values into the formula:

$$10{,}000 = 0 + (n-1)\cdot 50 + 0\cdot 0.003333^{(n-1)}$$

Simplify:

$$10{,}000 = (n-1)\cdot 50$$

Now, isolate (n):

$$n-1 = \frac{10{,}000}{50}$$

$$n-1 = 200$$

$$n = 200+1$$

$$n = 201$$

So, it will take 201 months for your savings to reach or exceed \$10,000.

This problem-solving example demonstrates the application of mixed sequences in real-life financial scenarios. By recognizing and combining the arithmetic (monthly deposit increment) and geometric (compounded interest) elements, we can efficiently determine the time required to achieve a financial goal. Such problems illustrate the versatility and practicality of mixed sequences in mathematical modeling and decision-making (Harder & Wilson, 1998; Hildebrandt et al., 2006). Hence, problem-solving with mixed sequences connects the mathematical interaction of arithmetic and geometric sequences, providing a useful approach to tackle a wide range of real-world challenges. By identifying and formulating mixed sequences, mathematicians and practitioners can gain deeper insights into complex phenomena and make informed decisions across various domains, including finance, demographics, epidemiology, and project management (Insani et al., 2021).

2.6. SEQUENCES IN ADVANCED MATHEMATICS

In advanced mathematics, sequences become a tool of profound complexity, offering insights into various mathematical realms, including analysis, number theory, and beyond. In this section, we will study advanced mathematics to understand the intricate roles that sequences play, particularly in the context of combining arithmetic and geometric elements, and their far-reaching implications (Levant, 2011).

2.6.1. Analyzing Convergence

In advanced mathematics, sequences are frequently scrutinized for their convergence properties. Convergence relates to the behavior of a sequence as it progresses toward a specific limit. This analysis is particularly relevant when dealing with mixed sequences that connect arithmetic and geometric components (Levant, 2011).

Consider a mixed sequence (a_n) that combines an arithmetic common difference (d) and a geometric common ratio (r). Mathematically, we can represent such a sequence as:

$$a_n = a_1 + (n-1) \cdot d + a_1 \cdot r^{(n-1)}$$

To assess the convergence of this sequence, we examine the behavior as (n) approaches infinity. This entails evaluating the limit as (n) tends to infinity:

$$\lim_{n \to ¥} a_n$$

Convergence analysis seeks to determine whether this limit exists and, if so, what value it converges to. The convergence properties can vary based on the values of (d) and

a. If $(|r|<1)$ (the geometric component converges to zero), and (d) is finite, the sequence (a_n) typically converges to a finite limit.

b. If $(|r| \geq 1)$ (the geometric component grows without bound), the sequence (a_n) may diverge (i.e., not converge to a finite limit) unless (d) is precisely tuned to offset the unbounded growth introduced by (r).

Understanding convergence is foundational in advanced mathematics, as it forms the basis for various mathematical analyzes, including calculus and real analysis.

2.6.2. Exploring Analytic Continuation

Sequences also intersect with the concept of analytic continuation, a powerful tool in complex analysis. Analytic continuation involves extending the domain of a function defined on a part of the complex plane to a larger region. Sequences come into play when dealing with analytic functions that exhibit specific sequences of values within their domains.

For instance, the Riemann zeta function, $\text{æ}(s)$ is an essential function in number theory and complex analysis. It is defined as an infinite series:

$$\text{æ}(s) = 1^{(-s)} + 2^{(-s)} + 3^{(-s)} + 4^{(-s)} + \ldots$$

This series converges for values of (s) with real parts greater than 1. However, when dealing with number theory, it becomes necessary to extend the function's domain to include values of (s) with real parts less than or equal to 1. Analytic continuation allows mathematicians to do this by finding another function that matches the original function's values within its domain of convergence while extending to a larger domain. Hence, sequences come into play when examining the values of the zeta function within its convergent domain. These values are meticulously analyzed to derive an analytic continuation that applies to a broader range of complex numbers.

2.6.3. Unraveling Chaos Theory

Advanced mathematics also discusses chaos theory, where sequences play a pivotal role in understanding deterministic chaos. Chaos theory explores seemingly random and unpredictable behavior that emerges from deterministic systems, often expressed through difference equations or iterated functions (Figure 2.8).

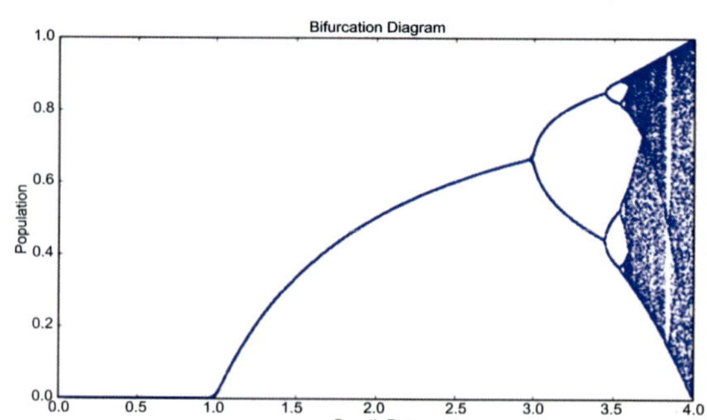

Figure 2.8. Bifurcation diagram showing chaos theory.

Source: Simon Vincek, Creative Commons License.

Consider the logistic map, a classic example of a discrete dynamical system. It is defined as:

$$x_{n+1} = r \cdot x_n \cdot (1 - x_n)$$

Here, (x_n) represents the value at a time (n), and (r) is a parameter. This iterative process generates a sequence of values, often starting with an initial value (x_0), which can exhibit highly complex behavior, including bifurcations, period-doubling, and chaotic regimes. The logistic map illustrates how simple recursive rules can give rise to sequences with intricate and seemingly chaotic behavior. The study of such sequences in chaos theory provides insights into phenomena like turbulence, weather patterns, and fractals, where deterministic systems can produce seemingly unpredictable outcomes (Peitgen et al., 1992).

2.6.4. Applications in Cryptography

Sequences, with their predictability and randomness properties, also find applications in cryptography. Cryptographic algorithms often rely on sequences to generate keys or pseudorandom numbers. These sequences must exhibit properties like unpredictability and uniform distribution to ensure the security of cryptographic systems. One such sequence is the Fibonacci sequence, which is both simple to compute and exhibits pseudorandom characteristics. It has been used in various cryptographic applications. In modern cryptography, advanced pseudorandom number generators (PRNGs) are designed to produce sequences that are computationally indistinguishable from truly random sequences. These PRNGs rely on intricate mathematical algorithms and properties of sequences to ensure security in data encryption and communication (Kowalek et al., 2019).

Did You Know?

Geometric sequences find applications in cryptography, particularly in the generation of secure keys. The ability to generate sequences with specific properties is essential for creating robust encryption algorithms.

2.7. EXERCISES AND PRACTICE PROBLEMS

To deepen our understanding of sequences and their applications, let's delve into a set of practical problems that involve various types of sequences.

These exercises will challenge your problem-solving skills and showcase how sequences can be applied in real-world scenarios (Grigorieva & Grigorieva, 2016).

2.7.1. Exercise 1: Arithmetic Sequence in Finance

You are planning for retirement and want to calculate how much you need to save each month to reach your financial goal. You have a target retirement fund of $1,000,000. Assuming an annual interest rate of 6% on your investments, how much should you save each month if you plan to retire in 30 years?

- **Solution:**

To calculate how much you need to save each month, you can use the formula for the future value of an annuity. In this case, it is an ordinary annuity because payments are made at the end of each period. The formula is:

$$FV = PMT \times \frac{(1+r)^n - 1}{r}$$

Where:

FV is the future value (your retirement fund goal) = $1,000,000.

PMT is the monthly payment (what you want to find).

r is the monthly interest rate $(6\%12 = 0.005)$.

n is the total number of periods $((30 \times 12 = 360)\, months)$.

Now, plug in the values and solve for (PMT):

$$1{,}000{,}000 = PMT \times \frac{(1+0.005)^{360} - 1}{0.005}$$

First, calculate the expression inside the parentheses:

$$(1.005)^{360} \approx 3.8697$$

Now, solve for (PMT):

$$1{,}000{,}000 = PMT \times \frac{3.8697 - 1}{0.005}$$

$$1{,}000{,}000 = PMT \times \frac{2.8697}{0.005}$$

Divide by 2.8697:

$$PMT \approx \frac{1,000,000}{2.8697} \approx 348,126.24$$

So, you need to save approximately $348,126.24 each month to reach your retirement goal of $1,000,000 in 30 years (Yıldız & Tüysüz, 2019).

2.7.2. Exercise 2: Geometric Sequence in Population Growth

You are studying a population of bacteria in a lab. Initially, there were 100 bacteria. Every hour, each bacterium divides into two, doubling the population. How many bacteria will there be after 24 hours?

Solution:

In this scenario, each bacterium doubles in number every hour, which forms a geometric sequence. To find out how many bacteria will be there after 24 hours, we can use the formula for the nth term of a geometric sequence:

$$a_n = a_1 \cdot r^{(n-1)}$$

Where:

a_n is the number of bacteria after (n) hours.

a_1 is the initial population = 100 bacteria.

r is the common ratio = 2 (each bacterium doubles).

n is the number of hours = 24.

Now, plug in the values and calculate:

$$a_{24} = 100 \cdot 2^{(24-1)}$$

$$a_{24} = 100 \cdot 2^{23}$$

$$a_{24} = 100 \cdot 8,388,608$$

$$a_{24} = 838,860,800$$

After 24 hours, there will be approximately 838,860,800 bacteria in the population.

2.7.3. Exercise 3: Mixed Sequence in Loan Payments

You have taken out a loan for $10,000 at an annual interest rate of 5%. You plan to make monthly payments. Initially, you pay $100 per month, and with each passing month, you increase your payment by \$10. How many months will it take to fully repay the loan?

Solution:

This problem involves a mixed sequence where the monthly payments form an arithmetic sequence, and the outstanding loan balance decreases in a geometric manner due to the interest (van Hijum, Vaughan, & Vogel, 2013). To find out how many months it takes to fully repay the loan, we can set up the following equations:

1. For the monthly payment sequence (a_n):
   i. a_1 = \$100 (initial monthly payment).
   ii. d = \$10 (monthly payment increment).
   iii. n = the number of months we want to find.
2. For the remaining loan balance sequence (L_n):
   i. L_0 = \$10,000 (initial loan amount).
   ii. $r = 1 + (0.05)/(12)$ (monthly interest rate).

We need to find the smallest (n) such that $(L_n \leq 0)$, indicating that the loan has been fully repaid. Now, let's set up the equations:

For (a_n):

$$a_n = a_1 + (n-1) \cdot d$$

For (L_n):

$$L_n = L_0 \cdot r^n$$

Now, we need to find the smallest (n) such that $(L_n \leq 0)$. Let's calculate it:

For (a_n):

$$a_n = 100 + (n-1) \cdot 10$$

For (L_n):

$$L_n = 10{,}000 \cdot \left(1 + \frac{0.05}{12}\right)^n$$

Now, set (L_n) to zero and solve for (n):

$$10,000 \cdot \left(1 + \frac{0.05}{12}\right)^n = 0$$

This equation has no real solutions because the loan balance will never reach zero with positive interest rates and monthly payments. This means you'll never fully repay the loan based on the given conditions (Xiao et al., 2017).

Activity 2.1

Imagine you are a financial advisor helping a client plan for their retirement. They have a choice between two investment options: one offering an annual interest rate of 6% (geometric sequence), and the other providing a fixed annual return of $5,000 (arithmetic sequence). How would you use your knowledge of arithmetic and geometric sequences to advise your client on the better investment choice for their retirement fund?

2.8. SUMMARY

In this chapter, a comprehensive exploration of arithmetic and geometric sequences is presented. The chapter commences by introducing arithmetic sequences, emphasizing the defining characteristic, which is the common difference between terms. The various aspects of arithmetic sequences are examined, including techniques for calculating specific terms, determining the nth term, and grasping the concepts of arithmetic mean and series associated with these sequences.

The discussion transitions to geometric sequences, where the focal point becomes the importance of the common ratio. Insights are provided into term calculations, the formulation of nth term expressions, and the elucidation of geometric means and series. The chapter further extends its examination to encompass mixed sequences, which seamlessly integrate both arithmetic and geometric elements. The versatility of transitioning between different sequence types is showcased, along with practical problem-solving scenarios involving mixed sequences and their real-world applications. To enhance comprehension, a range of exercises and practice problems covering arithmetic, geometric, and mixed sequences are presented. Detailed solutions and explanations accompany these exercises, offering readers a comprehensive understanding of the chapter's concepts.

REFERENCES

1. Aharonov, D., & Stephenson, K., (1998). Geometric sequences of discs in the apollonian packing. *St Petersburg Mathematical Journal C/C of Algebra I Analiz, 9*, 509–542.
2. Alexan, W., Medhat, H., Hamza, A., & Hussein, H., (2018). *Sequence-Based Bit-Cycling in Double Layer Message Security* (Vol. 1, No. 1, pp. 23–28). Paper presented at the 2018 Advances in Wireless and Optical Communications (RTUWO).
3. Alghar, M. Z., Susanti, E., & Marhayati, M., (2022). Ethnomathematics: Arithmetic sequence patterns of Minangkabau carving on singok gonjong. *JUPITEK: Jurnal Pendidikan Matematika, 5*(2), 145–152.
4. Amsari, D., Arnawa, I. M., & Yerizon, Y., (2022). Development of a local instructional theory for the sequences and series concept based on contextual teaching and learning. *Linguistics and Culture Review, 6*(2), 434–449.
5. Angraini, L. M., Larsari, V. N., Muhammad, I., & Kania, N., (2023). Generalizations and analogical reasoning of junior high school viewed from Bruner's learning theory. *Infinity Journal, 12*(2), 291–306.
6. Angstenberger, L., (2001). *Dynamic Fuzzy Pattern Recognition with Applications to Finance and Engineering* (Vol. 1, No. 1, pp. 1–19). Springer Science & Business Media.
7. Anjum, S., & Mukherjee, V., (2022). A novel arithmetic sequence pattern reconfiguration technique for line loss reduction of photovoltaic array under non-uniform irradiance. *Journal of Cleaner Production, 331*, 129822.
8. Arbib, M. A., & Manes, E. G., (1982). The pattern-of-calls expansion is the canonical fixpoint for recursive definitions. *Journal of the ACM (JACM), 29*(2), 577–602.
9. Arroyo, J., Espínola, R., & Maté, C., (2011). Different approaches to forecast interval time series: A comparison in finance. *Computational Economics, 37*, 169–191.
10. Barndorff-Nielsen, O. E., & Shephard, N., (2001). Non-Gaussian Ornstein–Uhlenbeck-based models and some of their uses in financial economics. *Journal of the Royal Statistical Society: Series B (Statistical Methodology), 63*(2), 167–241.

11. Bolker, B., & Pacala, S. W., (1997). Using moment equations to understand stochastically driven spatial pattern formation in ecological systems. *Theoretical Population Biology, 52*(3), 179–197.
12. Camouzis, E., & Kotsios, S., (2021). May's host–parasitoid geometric series model with a variable coefficient. *Results in Applied Mathematics, 11*, 100160.
13. Casey, B., Kersh, J. E., & Young, J. M., (2004). Storytelling sagas: An effective medium for teaching early childhood mathematics. *Early Childhood Research Quarterly, 19*(1), 167–172.
14. Chapman, S. T., Gotti, F., & Gotti, M., (2020). Factorization invariants of Puiseux monoids generated by geometric sequences. *Communications in Algebra, 48*(1), 380–396.
15. Darmayanti, R., Sugianto, R., & Muhammad, Y., (2022). Analysis of students' adaptive reasoning ability in solving HOTS problems arithmetic sequences and series in terms of learning style. *Numerical: Jurnal Matematika Dan Pendidikan Matematika,* 73–90.
16. DiDomenico, A. S., (1997). From fibonacci numbers to geometric sequences and the Binet formula by way of the golden ratio! *The Mathematics Teacher, 90*(5), 386–389.
17. Duan, S., & Babu, S., (2007). *Processing Forecasting Queries* (Vol. 1, No. 1, pp. 711–722). Paper presented at the Proceedings of the 33rd international conference on Very large data bases.
18. Dueñas-Osorio, L. A., (2005). *Interdependent Response of Networked Systems to Natural Hazards and Intentional Disruptions* (Vol. 1, No. 1, pp. 1–23). Georgia Institute of Technology.
19. Dwyer, M. B., Avrunin, G. S., & Corbett, J. C., (1998). *Property Specification Patterns for Finite-State Verification* (Vol. 1, pp. 7–15). Paper presented at the Proceedings of the second workshop on Formal methods in software practice.
20. Firdaus, A. M., Juniati, D., & Wijayanti, P., (2020). Number pattern generalization process by provincial mathematics Olympiad winner students. *Journal for the Education of Gifted Young Scientists, 8*(3), 991–1003.
21. Genkin, D., Valenta, L., & Yarom, Y., (2017). *May the Fourth be with you: A Microarchitectural Side Channel Attack on Several Real-World Applications of Curve 25519* (Vol. 1, pp. 845–855). Paper presented at the Proceedings of the 2017 ACM SIGSAC Conference on Computer and Communications Security.

22. Gong, Y., (2019). *Computational Image Understanding Incorporating Physics-based Modeling and Empirical Learning for Real-world Applications, 1*, 1–136.
23. Grassi, G., (2021). Chaos in the real world: Recent applications to communications, computing, distributed sensing, robotic motion, bio-impedance modeling, and encryption systems. *Symmetry, 13*(11), 2151.
24. Grigorieva, E., & Grigorieva, E., (2016). Real-life applications of geometric and arithmetic sequences. *Methods of Solving Sequence and Series Problems,* 191–225.
25. Gunnarsson, R., Ekdahl, A. L., Landén, J., & Tegnefur, J., (2016). Students' strategies to continue geometric number sequences. *ICT in Mathematics Education: The Future and the Realities, 109*, 117–160.
26. Gura, K., (1996). Growth and symmetry a mathematics course for liberal arts students. *Problems, Resources, and Issues in Mathematics Undergraduate Studies, 6*(4), 337–350.
27. Gutierrez, A., Benedicto, C., Jaime, A., & Arbona, E., (2018). The cognitive demand of a gifted student's answers to geometric pattern problems: Analysis of key moments in a pre-algebra teaching sequence. *Mathematical Creativity and Mathematical Giftedness: Enhancing Creative Capacities in Mathematically Promising Students,* 169–198.
28. Harder, L. D., & Wilson, W. G., (1998). Theoretical consequences of heterogeneous transport conditions for pollen dispersal by animals. *Ecology, 79*(8), 2789–2807.
29. Hildebrandt, K., Polthier, K., & Wardetzky, M., (2006). On the convergence of metric and geometric properties of polyhedral surfaces. *Geometriae Dedicata, 123*(1), 89–112.
30. Hingwala, D., Chatterjee, S., Kesavadas, C., Thomas, B., & Kapilamoorthy, T. R., (2011). Applications of 3D CISS sequence for problem solving in neuroimaging. *Indian Journal of Radiology and Imaging, 21*(02), 90–97.
31. Insani, K., Prihandoko, A. C., Sa'id, I. A., & Safik, M., (2021). Developing of learning tools based on science, technology, engineering, and mathematics (STEM) based on learning community to improve critical thinking ability in class X student's arithmetic sequences and arithmetic materials. In: *Journal of Physics: Conference Series* (Vol. 1839, No. 1, p. 012020). IOP Publishing.

32. Joan, D. R., (2015). Assortment of recreational mathematics an information for the future learner. *I-Manager's Journal on Mathematics, 4*(1), 1.
33. Johansen, A., & Sornette, D., (2001). Finite-time singularity in the dynamics of the world population, economic and financial indices. *Physica A: Statistical Mechanics and its Applications, 294*(3, 4), 465–502.
34. Jupri, A., Usdiyana, D., & Sispiyati, R., (2022). Introducing an arithmetic sequence using realistic mathematics education principles. In: *AIP Conference Proceedings* (Vol. 2575, No. 1). AIP Publishing.
35. Kanerva, P., (2009). Hyperdimensional computing: An introduction to computing in distributed representation with high-dimensional random vectors. *Cognitive Computation, 1*, 139–159.
36. Keim, D., Andrienko, G., Fekete, J. D., Görg, C., Kohlhammer, J., & Melançon, G., (2008). *Visual Analytics: Definition, Process, and Challenges* (Vol. 4950, pp. 154–175). Springer
37. Kendall, M. G., & Hill, A. B., (1953). The analysis of economic time-series-part i: Prices. *Journal of the Royal Statistical Society. Series A (General), 116*(1), 11–34.
38. Kondo, S., & Miura, T., (2010). Reaction-diffusion model as a framework for understanding biological pattern formation. *Science, 329*(5999), 1616–1620.
39. Kowalek, P., Loch-Olszewska, H., & Szwabiński, J., (2019). Classification of diffusion modes in single-particle tracking data: Feature-based versus deep-learning approach. *Physical Review E, 100*(3), 032410.
40. Levant, A., (2011). Finite-time stability and high relative degrees in sliding-mode control. In: *Sliding Modes After the First Decade of the 21st Century: State of the Art* (Vol. 412, pp. 59–92). Springer.
41. McEvoy, P. B., & Coombs, E. M., (1999). Biological control of plant invaders: Regional patterns, field experiments, and structured population models. *Ecological Applications, 9*(2), 387–401.
42. Morton, J. T., Sanders, J., Quinn, R. A., McDonald, D., Gonzalez, A., Vázquez-Baeza, Y., & Hyde, E. R., (2017). Balance trees reveal microbial niche differentiation. *MSystems, 2*(1), 1–11.
43. Na, S., Sun, W., Ingraham, M. D., & Yoon, H., (2017). Effects of spatial heterogeneity and material anisotropy on the fracture pattern and

macroscopic effective toughness of Mancos Shale in Brazilian tests. *Journal of Geophysical Research: Solid Earth, 122*(8), 6202–6230.

44. Ng, O. L., Leung, A., & Ye, H., (2023). Exploring computational thinking as a boundary object between mathematics and computer programming for STEM teaching and learning. *ZDM–Mathematics Education, 1*, 1–15.
45. Nixon, E. G., (2002). *An Investigation of the Influence of Visualization, Exploring Patterns and Generalization on Thinking Levels in the Formation of the Concepts of Sequences and Series* (Vol. 1, pp. 1–288). University of South Africa.
46. Ong, D. C., & Ponomarenko, V., (2008). The Frobenius number of geometric sequences. *Integers, 8*(1), 1–3.
47. Peitgen, H. O., Jürgens, H., Saupe, D., Peitgen, H. O., Jürgens, H., & Saupe, D., (1992). Length, area, and dimension: Measuring complexity and scaling properties. *Chaos and Fractals: New Frontiers of Science,* 183–228.
48. Pickover, C. A., (2011). *The Zen of Magic Squares, Circles, and Stars: An Exhibition of Surprising Structures Across Dimensions* (Vol. 1, pp. 2–19). Princeton University Press.
49. Qonita, Q., Prihatmanto, A. S., & Wuryandari, A. I., (2015). *Vidyanusa Game Utilization on Arithmetic Sequence and Addition Subtraction of Integers to Improve Mathematics Learning Outcomes of Junior High School Students (Case Study in SMPN 31 Bandung)* (Vol. 1, pp. 1–16). Paper presented at the 2015 4th International Conference on Interactive Digital Media (ICIDM).
50. Rachma, A. A., Rosjanuardi, R., & Suhendra, S., (2022). Students' mathematical obstacles in learning geometric sequences and series. In: *AIP Conference Proceedings* (Vol. 2468, No. 1). AIP Publishing.
51. Rahmani-Andebili, M., (2021a). Problems: Arithmetic and geometric sequences and series. In: *Precalculus: Practice Problems, Methods, and Solutions* (pp. 101–105). Springer.
52. Rahmani-Andebili, M., (2021b). Solutions of problems: Arithmetic and geometric sequences and series. In: *Precalculus: Practice Problems, Methods, and Solutions* (pp. 107–116). Springer.
53. Ram, M., (2020) De Gruyter series on the applications of mathematics in engineering and information sciences. *Techniques in Engineering Sciences* (Vol. 1, pp. 1–230). Berlin, Boston: De Gruyter.

54. Ross, A., (2017). A brief look at arithmetic and geometric sequences. In: *Pedagogy and Content in Middle and High School Mathematics* (Vol. 1, pp. 77–80). Brill.
55. Sari, M. S., & Susanti, E., (2020). Development of teaching materials arithmetic sequence and series based on android for problem-based learning. In: *Journal of Physics: Conference Series* (Vol. 1480, No. 1, p. 012024). IOP Publishing.
56. Sarker, I. H., (2021). Deep learning: A comprehensive overview on techniques, taxonomy, applications, and research directions. *SN Computer Science, 2*(6), 420.
57. Shiavi, R., (2010). *Introduction to Applied Statistical Signal Analysis: Guide to Biomedical and Electrical Engineering Applications* (Vol. 3, pp. 1–18). Elsevier.
58. Shih, S. Y., Sun, F. K., & Lee, H. Y., (2019). Temporal pattern attention for multivariate time series forecasting. *Machine Learning, 108*, 1421–1441.
59. Shute, V. J., Graf, E. A., & Hansen, E. G., (2006). Designing adaptive, diagnostic math assessments for individuals with and without visual disabilities. *ETS Research Report Series, 2006*(1), I–37.
60. Siagian, Q. A., Aswin, A., & Herman, T., (2023). Praxeological analysis of mathematics textbooks for class XI high school students on arithmetic and geometric sequences. *Eduma: Mathematics Education Learning and Teaching, 12*(2), 139–152.
61. Sun, H., Zhang, Y., Baleanu, D., Chen, W., & Chen, Y., (2018). A new collection of real-world applications of fractional calculus in science and engineering. *Communications in Nonlinear Science and Numerical Simulation, 64*, 213–231.
62. Tong, D. H., Uyen, B. P., Hai, Q. T., & Ngan, L. K., (2021). Experiential learning applied to the teaching of arithmetic and geometric sequences in high schools. *Multicultural Education, 7*(9), 1–19.
63. Ulam, S., (1962). On some mathematical problems connected with patterns of growth of figures. In: *Proceedings of Symposia in Applied Mathematics* (Vol. 14, No. 14, pp. 215–224). Providence, RI, USA: American Mathematical Society.
64. Vaidyanathan, P., (2014). Ramanujan sums in the context of signal processing—Part I: Fundamentals. *IEEE Transactions on Signal Processing, 62*(16), 4145–4157.

65. Van, H. S. A., Vaughan, E. E., & Vogel, R. F., (2013). Application of state-of-art sequencing technologies to indigenous food fermentations. *Current Opinion in Biotechnology, 24*(2), 178–186.
66. Wang, H., Liu, Y., Song, Y., Zhao, Y., Zhao, J., & Wang, D., (2012). Fractal analysis and its impact factors on pore structure of artificial cores based on the images obtained using magnetic resonance imaging. *Journal of Applied Geophysics, 86*, 70–81.
67. Xiao, X., Wang, Z., Li, Q., Xia, S., & Jiang, Y., (2017). Back-propagation neural network on Markov chains from system call sequences: A new approach for detecting Android malware with system call sequences. *IET Information Security, 11*(1), 8–15.
68. Yeshurun, S., (1978). The duality principle in teaching arithmetic and geometric series. *International Journal of Mathematical Education in Science and Technology, 9*(1), 65–70.
69. Yıldız, N., & Tüysüz, F., (2019). A hybrid multi-criteria decision-making approach for strategic retail location investment: Application to Turkish food retailing. *Socio-Economic Planning Sciences, 68*, 100619.
70. Yu, Q., Tang, H., Hu, J., Tan, C. K., Yu, Q., Tang, H., & Tan, K. C., (2017). Rapid feedforward computation by temporal encoding and learning with spiking neurons. *Neuromorphic Cognitive Systems: A Learning and Memory Centered Approach,* 19–41.
71. Zazkis, R., & Liljedahl, P., (2002). Generalization of patterns: The tension between algebraic thinking and algebraic notation. *Educational Studies in Mathematics, 49*, 379–402.
72. Zou, Y., Donner, R. V., Marwan, N., Donges, J. F., & Kurths, J., (2019). Complex network approaches to nonlinear time series analysis. *Physics Reports, 787*, 1–97.

CHAPTER 3

Fibonacci Numbers, the Golden Ratio, and the Laws of Nature

Contents

3.1. INTRODUCTION

The concepts of Fibonacci numbers and the Golden Ratio hold a unique position in mathematics, art, and the natural world. In this chapter, we will discuss Fibonacci numbers and the Golden Ratio, emphasized by precise mathematical definitions, historical context, and profound significance (MacQueen, 2006).

The origin of Fibonacci numbers can be traced back to the work of Leonardo of Pisa, an Italian mathematician also known as Fibonacci, in his book "Liber Abaci" published in 1202. Fibonacci introduced these numbers while contemplating a hypothetical population of rabbits, leading to the sequence's discovery. Their presence in this early work laid the foundation for further exploration and fascination among mathematicians (Ricketts, 1982).

The Golden Ratio, approximately equal to 1.6180339887, manifests itself when one divides a line into two segments in such a way that the ratio of the whole line to the longer segment equals the ratio of the longer segment to the shorter one. The Golden Ratio possesses remarkable mathematical properties, often appearing in art, architecture, and nature (Ozturk et al., 2016). The golden ratio can be found in the ratios of consecutive Fibonacci numbers.

As (n) approaches infinity, the ratio $\left(\frac{F_{n+1}}{F_n}\right)$ approaches φ. This relationship can be represented by the limit:

$$\mathop{\mathrm{Lim}}_{n \to ¥} \frac{F_{n+1}}{F_n} = f.$$

For instance, taking larger Fibonacci numbers, such as 144 and 233, their ratio is approximately equal to φ:

$$\frac{233}{144} \approx 1.6180555 \approx f.$$

The convergence of this ratio to the Golden Ratio as Fibonacci numbers increase in magnitude highlights the profound connection between these mathematical entities (Akhtaruzzaman & Shafie, 2011).

3.2. THE FIBONACCI SEQUENCE

The Fibonacci Sequence is a captivating numerical pattern that has intrigued mathematicians, scientists, and artists for centuries. In this section, we will discuss the Fibonacci sequence, from its fundamental definition to its remarkable mathematical properties and real-world applications (Akhtaruzzaman & Shafie, 2011).

The Fibonacci Sequence is an infinite sequence of numbers that begins with two initial terms: 0 and 1. Each subsequent term in the sequence is the sum of the two preceding terms. Mathematically, it can be defined as follows:

$$F(0)=0$$

$$F(1)=1$$

$$F(n)=F(n-1)+F(n-2) \text{ for } n \geq 2$$

Here, $F(n)$ represents the nth term of the sequence. The sequence starts as follows:

$$0,1,1,2,3,5,8,13,21,\ldots$$

The Fibonacci Sequence has captured the imagination of many due to its generation rule. As we progress, we will explore how this sequence unfolds through its recursive definition and mathematical representation (Kuliš & Hodžić, 2020) (Figure 3.1).

Figure 3.1. Each level of branching creates a Fibonacci number of arms.

Source: PEAKD, Creative Commons License.

3.2.1. Recursive Definition and Mathematical Representation

The recursive definition of the Fibonacci Sequence is of high significance. It captures the idea that each number in the sequence relies on the values of its two predecessors. Let's delve into this concept further (Alpago, 2022).

Consider $F(n)$, where (n) is any positive integer greater than or equal to 2. To calculate $F(n)$, we add the two preceding terms, $F(n-1)$ and $F(n-2)$. Using the formula below:

$$F(n) = F(n-1) + F(n-2)$$

This recursive formula generates the entire Fibonacci Sequence. Starting with $F(0) = 0$ and $F(1) = 1$, we can compute each subsequent term using this rule.

Let's illustrate this process with a few examples:

a. To find $(F(2))$, we add $F(1) + F(0)$, which equals $(1+0=1)$.

b. To find $(F(3))$, we add $F(2) + F(1)$, which equals $(1+1=2)$.

c. To find $(F(4))$, we add $F(3) + F(2)$, resulting in $(2+1=3)$.

d. Continuing this pattern, $F(5) = 5), (F(6) = 8$, and so on.

Mathematically, this recursive definition captures the essence of the Fibonacci Sequence, allowing us to generate each term by relying solely on the values of the preceding two terms. This simplicity and self-referential nature make the Fibonacci Sequence a remarkable mathematical phenomenon (Ghose & Bahadur, 2019).

3.2.2. Calculating Fibonacci Numbers

While the recursive approach is conceptually clear, it becomes computationally intensive for large values of (n) due to repeated calculations of the same terms.

One efficient method to calculate Fibonacci numbers is through matrix exponentiation. The Fibonacci numbers can be expressed as a matrix equation:

$$\begin{bmatrix} F(n) \\ F(n-1) \end{bmatrix} = \begin{bmatrix} 1 & 1 \\ 1 & 0 \end{bmatrix} \begin{bmatrix} F(n-1) \\ F(n-2) \end{bmatrix}$$

Here, the matrix $\begin{bmatrix} 1 & 1 \\ 1 & 0 \end{bmatrix}$ remains constant while we raise it to the power of $(n-1)$ to obtain $\begin{bmatrix} F(n) \\ F(n-1) \end{bmatrix}$. The Fibonacci number $F(n)$ is the first element of this resulting matrix.

Matrix exponentiation allows us to calculate Fibonacci numbers in logarithmic time complexity, making it an efficient method for large values of ( ) (Hoffman, 1999; Fanthorpe et al., 2013).

Let's illustrate this method with an example. Suppose we want to find $F(6)$:

We start with the initial matrix:

$$\begin{bmatrix} F(1) \\ F(0) \end{bmatrix} = \begin{bmatrix} 1 \\ 0 \end{bmatrix}$$

We then raise the matrix $\begin{bmatrix} 1 & 1 \\ 1 & 0 \end{bmatrix}$ to the power of 5 since $(n-1=6-1=5)$:

$$\begin{bmatrix} F(6) \\ F(5) \end{bmatrix} = \begin{bmatrix} 1 & 1 \\ 1 & 0 \end{bmatrix}^5 \begin{bmatrix} 1 \\ 0 \end{bmatrix}$$

Using matrix exponentiation techniques, we compute the result to be:

$$\begin{bmatrix} F(6) \\ F(5) \end{bmatrix} = \begin{bmatrix} 8 \\ 5 \end{bmatrix}$$

So, $F(6) = 8$.

Matrix exponentiation provides an efficient way to calculate Fibonacci numbers, particularly for large values of (n). It leverages the inherent mathematical structure of the sequence to yield results quickly and accurately (Folkerts, 2004).

3.2.3. Patterns and Properties of the Fibonacci Sequence

The Fibonacci Sequence is not just a list of numbers; it possesses various patterns and mathematical properties that have fascinated mathematicians for centuries. Let's explore some of these intriguing aspects.

3.2.4. Golden Ratio Approximation

The connection between the Fibonacci Sequence and the Golden Ratio (f) is one of the most captivating aspects of this sequence. As we explore larger values of (n) in the Fibonacci Sequence, an interesting pattern emerges (King, 2013) (Figure 3.2).

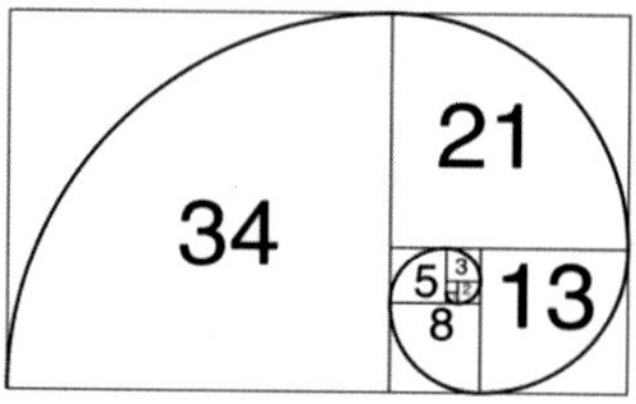

Figure 3.2. Golden rectangle with the first few terms from the Fibonacci sequence. Each number indicates the dimensions of the square it is in.

Source: StudyPug, Creative Commons License.

3.2.4.1. Fibonacci Spiral

One of the most remarkable manifestations of the Fibonacci Sequence is the Fibonacci Spiral. This spiral is constructed by drawing quarter circles with radii equal to consecutive Fibonacci numbers. The result is a captivating spiral pattern that appears in various natural phenomena and artworks. It is a representation of the mathematical order found in nature.

The Fibonacci Spiral is deeply linked to the Fibonacci Sequence itself. As one continues the sequence and creates larger quarter circles, the spiral's shape becomes more pronounced. This spiral is frequently observed in nature, from the arrangement of seeds in a sunflower to the growth patterns of certain shells. Its presence underscores the mathematical principles that govern the natural world (Debnath, 2011; Scott & Marketos, 2014) (Figure 3.3).

Figure 3.3. Fibonacci spirals found in nature.

Source: Adafruit, Creative Commons License.

3.2.4.2. Fibonacci Numbers in Nature

The prevalence of Fibonacci numbers in nature is a testament to their importance. One of the most well-known examples is the arrangement of leaves on a plant stem. In many plants, leaves are arranged in a spiral pattern, and the number of spirals often corresponds to Fibonacci numbers. This arrangement allows leaves to receive maximum sunlight exposure without shading each other. Fibonacci numbers also dictate the branching of trees. The number of branches at each level of a tree often follows the Fibonacci sequence. This pattern allows trees to efficiently capture sunlight for photosynthesis (Glushkov, 1976; Moyon, 2019) (Figure 3.4).

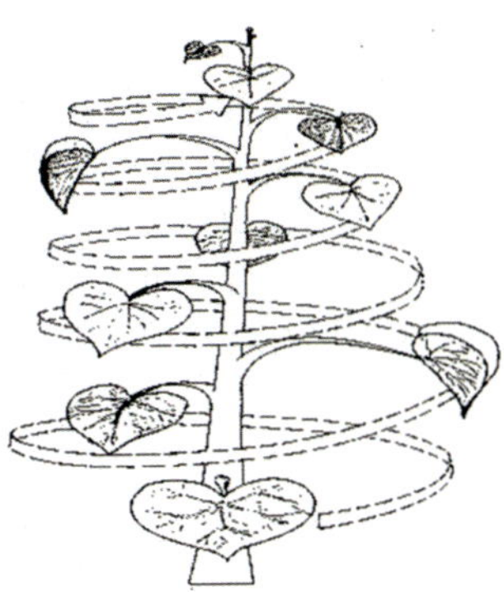

Figure 3.4. *The Fibonacci Series in the distribution of leaves around a central stem: 3 leaves in 5 turns, 5 leaves in 8 turns.*

Source: Mehrdad Hejazi, Creative Commons License.

Additionally, Fibonacci numbers can be found in the spirals of pinecones, pineapples, and even hurricanes. The presence of these numbers in natural phenomena showcases the fundamental role of mathematics in shaping the world around us (Benavoli et al., 2009; Parshall, 2017).

3.2.4.3. Nature's Optimization

The Fibonacci Sequence reflects nature's optimization. It represents an efficient way for plants to maximize their exposure to sunlight and minimize overcrowding. Through evolutionary processes, plants have developed this pattern to ensure their survival and growth. The arrangement of leaves and branches in a Fibonacci pattern allows each part of the plant to access sunlight effectively, leading to healthier and more robust growth (Watson, 2017).

This optimization extends beyond plants and is evident in various aspects of the natural world, from the spiral patterns of seashells to the growth patterns of certain animals. Nature's adoption of Fibonacci numbers as an optimizing principle is a testament to the elegance and efficiency of mathematical patterns (Figure 3.5).

Figure 3.5. Fibonacci spiral patterns of seashells.

Source: Polymath07, Creative Commons License.

3.2.4.4. Diverse Applications

Beyond their prevalence in the natural world, Fibonacci numbers find applications in diverse fields. In computer science and algorithms, Fibonacci numbers are used in various contexts.

They play a role in optimizing algorithms, calculating the greatest common divisors, and even generating pseudorandom numbers for cryptographic applications. In the world of finance, Fibonacci numbers are employed in technical analysis. Traders and analysts use Fibonacci retracement levels to identify potential support and resistance levels in financial markets. This application highlights the versatility of Fibonacci numbers in quantitative analysis (Winkelmann et al., 2006).

As we delve deeper into the Fibonacci Sequence, we will continue to uncover more mathematical insights, explore additional patterns, and dive into practical applications. These properties not only make the Fibonacci Sequence a captivating mathematical concept but also underscore its profound significance in mathematics, nature, and technology.

3.3. THE GOLDEN RATIO (f)

The Golden Ratio, symbolized by the Greek letter (f), is an interesting mathematical constant that goes beyond disciplines. At its core, it represents a unique proportion in which a line is divided into two parts. The ratio of the whole line to the longer segment is the same as the ratio of the longer segment to the shorter segment. This ratio, (f), approximately equals 1.6180339887.

Mathematically, if we have a line segment (AB) where (A) divides the segment into two parts, (AC) and (CB), such that:

$$\frac{AB}{AC} = \frac{AC}{CB} = f$$

Then, (f) is the Golden Ratio. This division point (A) results in a proportion that is not only mathematically interesting, which is why it has fascinated artists and architects throughout history (Salingaros, 2012).

3.3.1. Expressing (f) Mathematically

The Golden Ratio can be expressed mathematically as the positive solution to the quadratic equation:

$$x^2 - x - 1 = 0$$

Solving this equation yields the value of (f):

$$x = \frac{1+\sqrt{5}}{2}$$

This expression encapsulates the essence of (f) and provides an exact representation of this intriguing constant. Approximated as 1.6180339887, (f) reveals its intricacy through its irrationality, emphasizing the depth of its mathematical significance (Omotehinwa & Ramon, 2013).

Another mathematical representation of (f) is through a continued fraction:

$$\Phi = 1 + \cfrac{1}{1 + \cfrac{1}{1 + \cfrac{1}{1 + \cfrac{1}{1 + \cfrac{1}{1 + \cdots}}}}}$$

This continued fraction showcases the recursive nature of the Golden Ratio, where it can be approximated through a series of unit fractions. It offers a unique perspective on how (f) can be iteratively approached (Sigalotti & Mejias, 2006).

3.3.2. The Relationship Between Consecutive Fibonacci Numbers and (f)

The Fibonacci Sequence, characterized by its recursive nature, generates a sequence of numbers where each term is the sum of the two preceding terms. As we explore larger values of (n) in the Fibonacci Sequence, the ratio of consecutive Fibonacci numbers $F(n)/F(n-1)$ approaches the Golden Ratio (f):

$$\operatorname*{Lim}_{n \to ¥} \frac{F(n)}{F(n-1)} = f$$

This convergence to the Golden Ratio occurs as (n) becomes larger. It reveals the inherent connection between the Fibonacci Sequence and the Golden Ratio. This relationship is an indication of the mathematical smartness of both concepts (Fauvel et al., 2006). To further illustrate this relationship, consider the following examples:

a. For $(n=5)$, we have $F(5)=5$ and $F(4)=3$. The ratio $F(5)/F(4)$ is $(5/3)$, which approximates (f).

b. For $(n=8)$, we have $F(8)=21$ and $F(7)=13$. The ratio $F(8)/F(7)$ is $(21/13)$, which is a closer approximation of (f).

c. As we continue this pattern with larger values of (n), the ratio $F(n)/F(n-1)$ approaches the Golden Ratio more closely.

This remarkable property highlights how the Fibonacci sequence naturally encapsulates the Golden Ratio within its structure, making it a powerful tool for approximating this irrational number (Iosa et al., 2013; Tahir, 2023).

3.3.3. Geometric Constructions of the Golden Ratio

The presence of the Golden Ratio extends beyond mathematical expressions and into geometric constructions, offering captivating visual representations of this mathematical constant. Two notable geometric constructions involving the Golden Ratio are the Golden Rectangle and the Golden Spiral (Eirgash & Toğan, 2023).

3.3.3.1. The Golden Rectangle

A Golden Rectangle is a rectangular shape in which the ratio of its longer side to its shorter side is the Golden Ratio (f). If we denote the length of the longer side as (a) and the length of the shorter side as (b), then:

$$\frac{a}{b}=f$$

This proportion defines the unique shape of the Golden Rectangle, characterized by its visually appealing proportions. It possesses a captivating property: if you remove a square from one of the corners of a Golden Rectangle, the remaining portion is also a Golden Rectangle. This process can be repeated infinitely, creating a visually pleasing spiral known as the Golden Spiral (Kazankov & Gubin, 2020) (Figure 3.6).

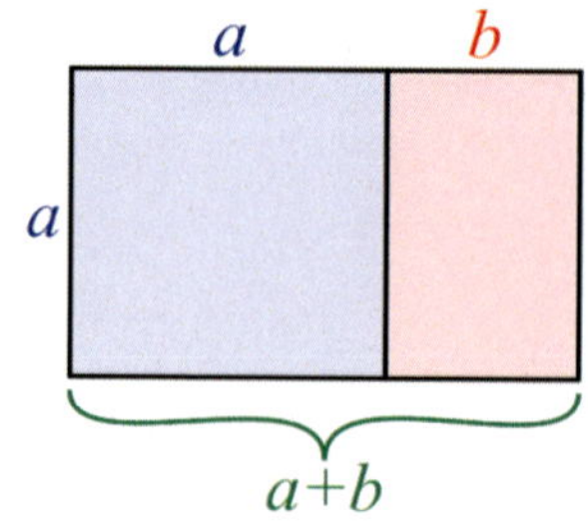

Figure 3.6. A golden rectangle with sides ab placed adjacent to a square with sides of length a produces a similar golden rectangle.

Source: Ahecht, Creative Commons License.

3.3.3.2. The Golden Spiral

The Golden Spiral is a self-replicating spiral that appears in various natural phenomena, from seashells to galaxies. It is constructed using a series of Golden Rectangles, each successively smaller than the previous one. The curvature of the spiral closely approximates the Golden Ratio (Zeina & Ahmed, 2022). To construct a Golden Spiral, you start with a square. Then, using the Golden Ratio, you draw a smaller square adjacent to it. This process is repeated, with each square's side length determined by the Golden Ratio relative to the previous square. By connecting the corners of these squares with smooth curves, the Golden Spiral emerges (Nematollahiet al., 2020).

3.3.3.3. The Golden Ratio in Art and Architecture

The aesthetic aspect of the Golden Ratio has not gone unnoticed by artists and architects. Throughout history, it has been employed to create visually pleasing compositions and designs. In art, the Golden Ratio has been used to determine the proportions of paintings, sculptures, and even musical compositions. It is believed that art that adheres to the Golden Ratio is more aesthetically pleasing to the human eye (Verrelli et al., 2023).

In architecture, the Golden Ratio has been used to design buildings, facades, and interiors. Architects have incorporated the Golden Ratio's proportions into their designs to create structures that are both functional and visually harmonious. Famous examples include the Parthenon in Athens, which exhibits Golden Ratio proportions in its columns and pediments (Rizzo, 2023) (Figure 3.7).

Figure 3.7. Employment of the phi in the great pyramid of Giza.

Source: Illustrarch, Creative Commons License.

3.3.3.4. The Golden Ratio in Nature

The presence of the Golden Ratio in the natural world is a testament to its significance. In plants, the arrangement of leaves and branches often follows the Fibonacci sequence and, by extension, exhibits Golden Ratio proportions. This arrangement allows plants to efficiently capture sunlight and maximize their growth potential. The spirals found in shells, such as the nautilus, also exhibit the Golden Ratio. The nautilus shell's chambers are arranged in a logarithmic spiral, a manifestation of the Golden Ratio's prevalence in the natural world (Verrelli et al., 2021). Additionally, the proportions of the human body, such as the length of the arm relative to the forearm, have been found to approximate the Golden Ratio. This ratio has even been suggested to influence perceptions of beauty and symmetry (Figure 3.8).

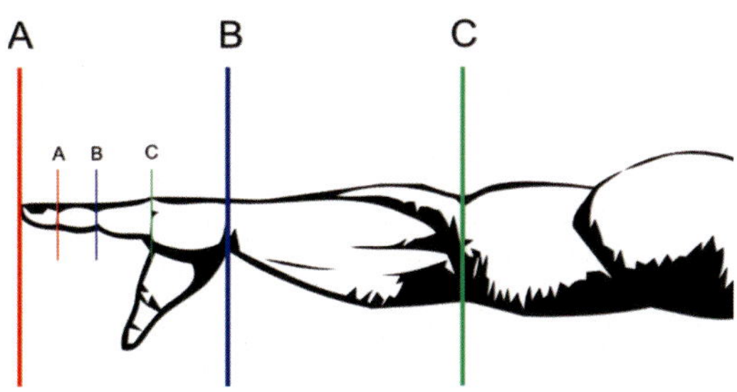

Figure 3.8. Your hand creates a golden section in relation to your arm, as the ratio of your forearm to your hand is also 1.618, the divine proportion.

Source: The Golden Ratio, Creative Commons License.

3.3.3.5. The Golden Ratio in Modern Science and Technology

The Golden Ratio's influence extends into modern science and technology. In mathematics and number theory, it continues to be a subject of research. Its unique properties make it a rich field for exploration, with applications in various mathematical concepts and algorithms. In computer science, the Golden Ratio is used in algorithms for generating pseudorandom numbers, which are crucial for cryptography and secure communication. Moreover, in design and user experience, the Golden Ratio is sometimes employed to create esthetically pleasing interfaces and layouts for websites and applications (Knill & Tangerman, 2011). Hence, unique properties, mathematical representations, and geometric constructions of the Golden Ratio (f) make it a subject of significance. From art and architecture to the natural world and modern technology, the Golden Ratio leaves its permanent mark on our perception of beauty, harmony, and mathematical phenomena.

3.4. APPLICATIONS OF FIBONACCI NUMBERS AND THE GOLDEN RATIO

The reach of Fibonacci numbers and the Golden Ratio extends far beyond the boundaries of mathematics and into diverse domains. In this section, we explore the profound impact of these mathematical concepts on various aspects of our world. From the intricate patterns found in nature, such as phyllotaxis and branching, to their presence in architecture, financial markets, and the algorithms that power our digital age, we will delve into the multifaceted applications of Fibonacci numbers and the Golden Ratio (Zheng & Bhunia, 2015). These applications showcase the enduring relevance and pervasive beauty of these mathematical wonders in our daily lives.

Remember

Fibonacci numbers appear widely in nature, from the arrangement of leaves on a stem to the spiral patterns in seashells. Understanding these natural occurrences can deepen your appreciation for mathematics in the real world.

3.4.1. Phyllotaxis

Consider a plant with leaves or petals arranged in a spiral pattern as they grow along the stem (Martinez et al., 2016). If we count the number of spirals in one direction and the number of spirals in the opposite direction, we frequently find consecutive Fibonacci numbers. For example, you might observe 21 spirals in one direction and 34 spirals in the other, mirroring successive Fibonacci numbers (Figure 3.9).

a. **Example 1:** Take the example of a sunflower. By counting the spirals in each direction, you might find 21 spirals in one direction and 34 in the other— numbers closely resembling the Fibonacci sequence.

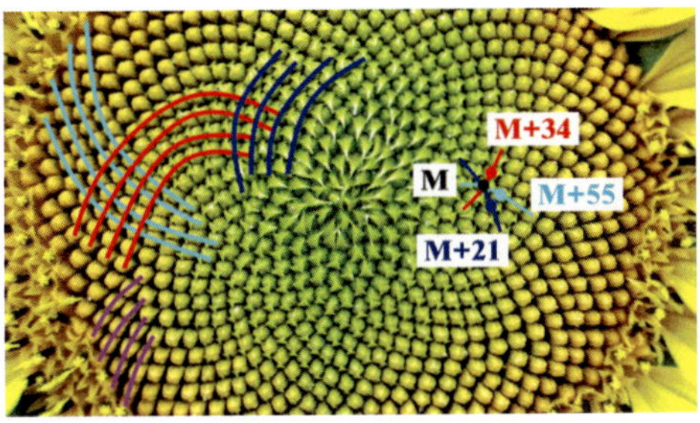

Figure 3.9. The phyllotactic pattern on the inflorescence of a sunflower.

Source: Boris Rozin, Creative Commons License.

b. **Example 2:** Pinecones exhibit a similar phenomenon, where the spirals housing the pinecone's scales often follow Fibonacci numbers, such as 5 spirals in one direction and 8 in the other.

The prevalence of Fibonacci numbers in phyllotaxis is not arbitrary; it is a result of nature's optimization for efficient packing. This arrangement ensures that each part, whether it be a leaf, petal, or seed, can access sunlight without shading others, maximizing the plant's photosynthesis and growth potential (Prusinkiewicz et al., 2022) (Figure 3.10).

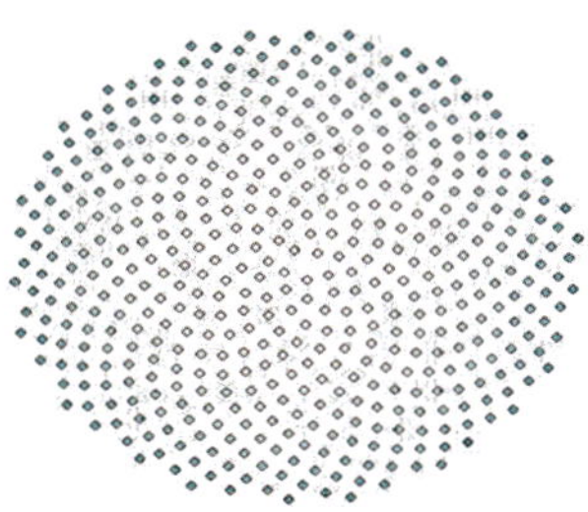

Figure 3.10. Depiction of spiral phyllotaxis.

Source: Mathrecreation, Creative Commons License.

3.4.1.1. Branching Patterns of Trees and Shrubs

Fibonacci numbers also exert their influence on the branching patterns of trees and shrubs. The number of branches and their arrangement frequently adhere to the Fibonacci sequence. Consider the branching of a tree as it grows. It bifurcates into branches, and each branch may further divide into smaller branches. This hierarchical branching system often exhibits a Fibonacci pattern. In many trees, the number of branches at each level follows the Fibonacci sequence. For instance, at the base of a tree, there may be one primary trunk (level 0). As you move up the tree, it bifurcates into two branches (level 1). These two branches may then divide into three branches each (level 2), and so on.

a. **Example 1**: The silver birch tree (*Betula pendula*) is known for its distinctive branching pattern. At each level, it often exhibits a Fibonacci number of branches, such as 1, 2, 3, 5, or 8, at different levels.

b. **Example 2:** Shrubs like the red-osier dogwood (*Cornus sericea*) and larger trees like the oak (genus Quercus) also demonstrate similar branching patterns.

The captivating aspect of this phenomenon is not merely aesthetic; it serves functional purposes. Each branching level optimizes light exposure for the leaves or photosynthetic structures on those branches, ensuring efficient energy capture (Yin & Kitazawa, 2021).

3.4.1.2. Fibonacci in Branch Lengths

In addition to the number of branches, Fibonacci ratios also manifest in the lengths of branches. The length of one branch relative to another often approximates the Golden Ratio (f).

The ratio of the length of a branch to its parent branch frequently approximates the Golden Ratio. This ratio is visually appealing and contributes to the overall structural stability and aesthetics of the tree or shrub.

a. **Example 1:** In pine tree branching, the ratio of a smaller branch's length to its parent branch often approximates the Golden Ratio (f).

b. **Example 2:** Silver birch trees, with their Fibonacci branching, also showcase this ratio in branch lengths.

3.4.1.3. The Parthenon: Architectural Brilliance

The Parthenon, an ancient Greek temple, is a marvel of architectural precision. Its base dimensions form a Golden Rectangle, where the ratio of length (L) to width (W) equals (f). This proportionality is expressed mathematically as:

Figure 3.11. Depiction of the Parthenon.

Source: Project Rhea, Creative Commons License.

$$\frac{L}{W} = f$$

Additionally, the temple's façade conforms to a logarithmic spiral reminiscent of the Golden Spiral. This spiral aligns with the Golden Ratio, contributing to the Parthenon's visual harmony (Sadoc & Charvolin, 2012; Yin & Kitazawa, 2021) (Figure 3.11).

3.4.1.4. The Mona Lisa: Enigmatic Composition

Leonardo da Vinci's Mona Lisa is renowned for its captivating composition. Its dimensions create a Golden Rectangle, with the ratio of height (H) to width (W) approximating (f):

$$\frac{H}{W} \approx f$$

This proportion enhances the painting's balance. Moreover, Leonardo's strategic placement of the subject adheres to a logarithmic spiral resembling the Golden Spiral, directing the viewer's focus and enriching the artwork's attraction (Meicenheimer & Zagorska-Marek, 1989) (Figure 3.12).

Figure 3.12. Image of Mona Lisa.

Source: Huntley, Creative Commons License.

In both cases, the Golden Ratio's mathematical precision underlies the enduring attraction of these arts, showcasing the profound interplay between mathematics and aesthetics in art and architecture (Rozin, 2020).

3.4.2. Fibonacci Analysis and the Financial Markets

Fibonacci analysis is a powerful tool in financial markets, allowing traders and investors to identify potential price levels, trends, and reversals. Derived from the Fibonacci sequence, this mathematical concept reveals hidden patterns within market price movements.

3.4.2.1. Fibonacci Retracement Levels

One of the central elements of Fibonacci analysis in financial markets is the identification of retracement levels. These levels help traders pinpoint potential areas of support and resistance, aiding in entry and exit strategies. The key retracement levels are derived from the Fibonacci sequence and expressed as percentages:

a. **38.2%:** This level is not a true Fibonacci number but is included due to its popularity. It often serves as an initial level of support or resistance.

b. **50.0%:** While not a Fibonacci number, this level is significant because it represents a simple 50% retracement, indicating a potential reversal or continuation point.

c. **61.8%:** Known as the "Golden Ratio" in financial markets, this level often serves as a strong level of support or resistance. It is derived directly from the Fibonacci sequence $\left(\frac{1}{f} \approx 0.618\right)$, making it a key focal point for traders.

3.4.2.2. Fibonacci Extensions

Fibonacci extensions are used to project potential future price levels beyond the current trend. These extensions are also based on Fibonacci ratios and are commonly used by traders to set profit targets. The primary Fibonacci extensions are:

a. **161.8%:** This extension level is derived from the Golden Ratio (f), making it a significant point to watch for potential price continuation.

b. **261.8%:** Double the Golden Ratio; this extension level indicates an even stronger potential for price extension.

c. **423.6%:** Derived from the Fibonacci sequence, this level suggests a substantial extension beyond the initial trend.

3.4.2.3. Fibonacci Fan Lines

Fibonacci fan lines are a valuable tool for traders to identify potential trendlines and trend reversals. These fan lines are drawn based on specific Fibonacci ratios and provide insights into potential areas of support and resistance (Figure 3.13).

The primary fan lines include:

a. **38.2% Fan Line:** This fan line is derived from the 38.2% retracement level and can act as a support or resistance line.

b. **50.0% Fan Line:** Corresponding to the 50% retracement level, this fan line is another potential trendline.

c. **61.8% Fan Line:** This fan line is aligned with the Golden Ratio and often represents a strong trendline in financial markets.

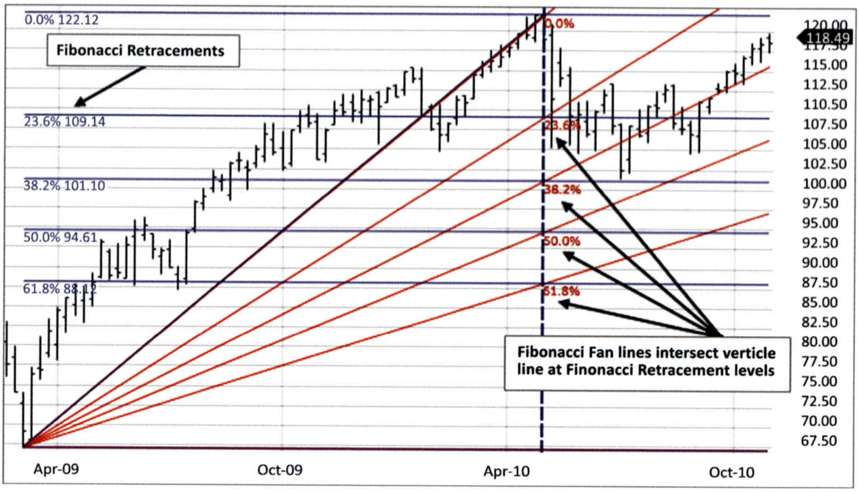

Figure 3.13. Graphical representation of Fibonacci fan lines.

Source: Fidelity, Creative Commons License.

3.4.2.4. Steps in Fibonacci Analysis

To illustrate the practical application of Fibonacci analysis, consider a scenario in which a trader observes a strong upward trend in a stock's price. The trader believes that the stock may experience a retracement before continuing its upward movement. The following steps are used in Fibonacci analysis (Drăgoi, 2017; Williams, 2018):

1. **Identifying the Initial Trend:** The trader identifies the initial trend and measures the price movement from a significant low to a high point.
2. **Drawing Fibonacci Retracement Levels:** Using the Fibonacci retracement tool, the trader draws retracement levels on the price chart. These levels are typically drawn from the low point to the high point of the trend.
3. **Assessing Potential Support Levels:** The trader identifies potential support levels based on the Fibonacci retracement levels, with a particular focus on the 38.2%, 50.0%, and 61.8% levels.
4. **Monitoring Price Action:** As the price retraces, the trader closely monitors the price action around the Fibonacci retracement levels.

If the price shows signs of support or reverses at one of these levels, it may indicate a potential buying opportunity.

5. **Setting Profit Targets:** To set profit targets, the trader can use Fibonacci extensions. These levels are drawn from the initial low to the high point of the trend and can help identify potential price targets for the upward movement.
6. **Managing Risk:** Effective risk management is crucial in trading. Traders often use stop-loss orders to limit potential losses if the market moves against their position.

3.4.2.5. Example of Stock Price

Suppose a trader identifies an upward trend in a stock's price, with the low point at $100 and the high point at $150. Using Fibonacci retracement levels, the trader draws lines at the 38.2%, 50.0%, and 61.8% levels, which correspond to $122, $125, and $128, respectively.

If the stock's price retraces and finds support at the $125 level, the trader may interpret this as a potential buying opportunity, expecting the stock to continue its upward movement (Minarova, 2014).

To set profit targets, the trader uses Fibonacci extensions. Drawing the extension levels from the initial low of $100 to the high of $150, the trader identifies potential price targets at the 161.8% ($167.80), 261.8% ($182.20), and 423.6% ($204.36) levels.

By incorporating Fibonacci analysis, traders can make informed decisions based on mathematical patterns and ratios, enhancing their ability to navigate the complexities of financial markets. It serves as a valuable tool for technical analysis, offering insights into potential price levels, trendlines, and reversal points (Minarova, 2014).

3.4.3. Applications in Computer Science and Algorithms

The influence of Fibonacci numbers and the Golden Ratio ( ) extends beyond the realms of mathematics and into the domain of computer science and algorithms. These mathematical concepts find practical applications in various computational fields, offering solutions to problems and enhancing algorithmic efficiency (Jin et al., 2021).

3.4.3.1. Fibonacci Numbers in Computer Science

Fibonacci numbers have a direct role in computer science as a classic sequence generation problem. In computer science, this recursive definition can be implemented using algorithms such as recursive functions or iterative loops. Generating Fibonacci numbers is used in various applications, including mathematical modeling and cryptography (Levitov & Letters, 1991).

Dynamic programming is a powerful algorithmic technique that uses problem decomposition and memorization to solve complex problems efficiently. Fibonacci numbers provide a classic example of dynamic programming's utility. Instead of repeatedly recalculating Fibonacci numbers, dynamic programming stores intermediate results, dramatically reducing computation time.

The Fibonacci problem can be solved using dynamic programming in $O(n)$ time, where (n) is the desired Fibonacci number. This efficiency makes dynamic programming an essential tool in computer science, particularly in algorithm optimization (Goldstine, 2006).

3.4.3.2. The Golden Ratio in Computer Science

The golden ratio plays a significant role in the following key areas of computer science:

- **Algorithm Efficiency:** The Golden Ratio (f) and its reciprocal $\left(\frac{1}{f}\right)$ often appear in algorithms and data structures to optimize performance. For example, the use of the Golden Ratio in hash functions can lead to more even data distribution and reduced collisions. This concept is vital in databases, where efficient data retrieval is essential.
- **Search Algorithms:** The Golden Ratio influences the design of search algorithms. The Fibonacci search, inspired by Fibonacci numbers, is one such algorithm. It efficiently searches for a target value in a sorted array by dividing the search interval according to the Golden Ratio (Wahl, 2023).

The formula for dividing the interval in the Fibonacci search is:

$$k = \frac{F(n-2)}{F(n)}$$

Where $F(n)$ is the Fibonacci number closest to but not exceeding the array size. This method allows for a balanced search and often requires fewer comparisons than binary search in certain scenarios.

3.4.3.3. Combining Fibonacci Numbers and the Golden Ratio

Their combination is helpful in the following areas:

- **Algorithm Analysis:** The mathematical properties of Fibonacci numbers and the Golden Ratio can assist in algorithm analysis. For example, in analyzing the efficiency of recursive algorithms that use the Fibonacci sequence, the recursive call stack's depth can be expressed using Fibonacci numbers. Understanding this depth aids in predicting an algorithm's performance.
- **Random Number Generation:** The Golden Ratio plays a role in pseudo-random number generation. The golden ratio (f) is an irrational number, which means it has a non-repeating, non-terminating decimal expansion. This property makes it useful in generating sequences of numbers that exhibit pseudorandom behavior, a crucial element in simulations, cryptography, and computer graphics.

3.4.3.4. Case Study: Fibonacci Heap in Data Structures

The Fibonacci heap is a data structure used in computer science for priority queue operations, such as Dijkstra's algorithm and Prim's algorithm for minimum spanning tree construction. It employs Fibonacci numbers to achieve amortized constant-time operations.

The key attributes of a Fibonacci heap include:

1. **Constant-Time Minimum:** A Fibonacci heap can find and remove the minimum element in constant time, $O(1)$, which is advantageous for algorithms that require frequent minimum extractions.

2. **Amortized Constant-Time Operations:** The Fibonacci heap achieves amortized constant-time insertions and decrease-key operations. The amortized analysis, often involving Fibonacci numbers, ensures that the average cost of these operations remains constant over a sequence of operations.
3. **Lazy Merging:** The data structure uses lazy merging to consolidate heaps efficiently. It combines heaps of equal sizes to maintain performance. Fibonacci numbers guide the structure of the consolidation process.

- **Example of Fibonacci Heap:** Consider a scenario where Dijkstra's algorithm is used to find the shortest path in a graph with a large number of vertices. The efficiency of this algorithm is significantly improved when a Fibonacci heap is employed for priority queue operations. The constant-time minimum extraction and amortized constant-time decrease-key operations contribute to a faster runtime. Fibonacci numbers and the Golden Ratio serve as more than just mathematical curiosities; they are indispensable tools in the world of computer science and algorithms. From sequence generation and dynamic programming to algorithm efficiency and data structures like the Fibonacci heap, these mathematical concepts play a crucial role in enhancing computational efficiency and problem-solving capabilities. Harnessing the inherent feature of Fibonacci numbers and the Golden Ratio, computer scientists and algorithm designers continue to unlock innovative solutions to complex computational challenges.

3.5. MATHEMATICAL RELATIONSHIPS

The world of mathematics is rife with captivating relationships and properties, and within it, the Fibonacci sequence and the Golden Ratio hold a special place. In this section, we will explore the intricate connections and properties that surround these two mathematical concepts. We begin by discussing the connection between the Fibonacci sequence and Binet's formula, a remarkable expression that enables us to compute Fibonacci numbers efficiently. Next, we will discuss Cassini's identity and other mathematical properties that reveal the hidden patterns within the Fibonacci sequence. Finally, we will study continued fractions, where we witness the convergence to the Golden Ratio, uncovering the mathematical bonds that link these concepts together. (Atela et al., 2003).

3.5.1. The Fibonacci Sequence and Binet's Formula

Binet's Formula stands as a mathematical concept, offering a direct route to computing any term within the Fibonacci sequence without the need for extensive iterative calculations. This formula, attributed to the French mathematician Jacques Philippe Marie Binet, provides a concise closed-form expression:

$$F(n) = \frac{f^{n} - (-f)^{-n}}{\sqrt{5}}$$

Binet's Formula, a mathematical tool, not only simplifies the calculation of Fibonacci numbers but also reveals useful insights into the nature of this sequence.

- **Binet's Formula:** To study the practicality of Binet's Formula, let's consider an example.

Calculating the 7th Fibonacci number, denoted as $F(7)$.
Using Binet's Formula, we embark on a mathematical journey:

$$F(7) = \frac{f^{7} - (-f)^{-7}}{\sqrt{5}}$$

By substituting the approximate value of $(f)(1.6180339887)$, we solve the equation:

$$F(7) \approx \frac{1.6180339887^{7} - (-1.6180339887)^{-7}}{\sqrt{5}}$$

After conducting the mathematical study, we arrive at the result: $(F(7) \approx 13)$.

Binet's Formula encapsulates mathematical beauty in several ways:

a. **The Golden Ratio** (f)**: The** presence of the Golden Ratio in Binet's Formula makes it even more special. The formula's foundation lies in the Fibonacci sequence, inherently linked to the Golden Rati $= 1 + —)$. This mathematical relationship links the Fibonacci sequence and the Golden Ratio.

b. **Negative Powers of (f):** Binet's Formula deals with both (f^n) and $(-f)^{-n}$). Surprisingly, $(-f)^{-n}$) equates to $\left(\frac{1}{f^n}\right)$, creating a symmetrical balance within the formula. This symmetry mirrors the alternating nature of Fibonacci numbers, where even-indexed terms are positive and odd-indexed terms are negative.

c. **The Role of the Square Root of 5:** The denominator $(\sqrt{5})$ in Binet's Formula serves as a scaling factor, ensuring that the result remains an integer. Without it, the formula would yield irrational numbers, which would not align with the essence of the Fibonacci sequence.

Binet's Formula not only offers practicality but also serves as a gateway to deeper insights into the Fibonacci sequence (Kim et al., 2008).

3.5.2. Cassini's Identity and Other Mathematical Properties

Cassini's Identity is a remarkable mathematical relationship that highlights the properties of the Fibonacci sequence. While it may not be as well-known as Binet's Formula, it offers unique insights into the interplay between consecutive Fibonacci numbers and adds depth to our understanding of this sequence.

Cassini's Identity is expressed as follows:

$$F(n+1) \cdot F(n-1) - (F(n))^2 = (-1)^n$$

In this formula, $F(n)$ represents the nth term of the Fibonacci sequence. This identity highlights the intriguing and alternating behavior of Fibonacci numbers and serves as a testament to their mathematical richness (Stoma et al., 2008).

Mathematical Insights from Cassini's Identity:

a. **Alternating Behavior:** Cassini's Identity reveals the alternating nature of Fibonacci numbers. The right-hand side of the equation, $(-1)^n$, produces alternating signs (positive for even n and negative for odd n). This mirrors the alternating behavior of Fibonacci numbers themselves.

b. **Product of Consecutive Terms:** The left-hand side of the equation involves the product of consecutive Fibonacci terms, $F(n+1)\cdot F(n-1)$, and the square of $F(n)$. This product relationship encapsulates a fundamental property of the Fibonacci sequence.

c. **Significance of n:** The value of (n) plays a pivotal role in determining the sign on the right-hand side. When (n) is even, the right-hand side evaluates to 1, and when (n) is odd, it evaluates to -1. This alternating sign reflects the oscillation between positive and negative Fibonacci numbers.

Let's delve deeper into the implications of Cassini's Identity:

- **Case 1: Even n**

When (n) is even, the right-hand side evaluates to 1. This means that the product of consecutive Fibonacci terms, $F(n+1)\cdot F(n-1)$, exceeds the square of $F(n)$ by exactly 1. This behavior emphasizes the "growing" aspect of even-indexed Fibonacci numbers.

For example, consider $(n=4)$:

$$F(4+1)\cdot F(4-1)-(F(4))^2 = F(5)\cdot F(3)-(F(4))^2 = 5\cdot 2-3^2 = 10-9 = 1$$

- **Case 2: Odd n**

When (n) is odd, the right-hand side evaluates to -1. In this case, the product of consecutive Fibonacci terms, $F(n+1)\cdot F(n-1)$, falls short of the square of $F(n)$ by 1. This reflects the "contracting" nature of odd-indexed Fibonacci numbers.

For example, consider $(n=5)$:

$$F(5+1)\cdot F(5-1)-(F(5))^2 = F(6)\cdot F(4)-(F(5))^2 = 8\cdot 3-5^2 = 24-25 = -1$$

Cassini's Identity, while seemingly simple, encapsulates the essence of the Fibonacci sequence's behavior. It showcases how the product of consecutive terms relates to the square of the current term and how this relationship alternates depending on whether (n) is even or odd.

This mathematical identity is a testament to the depth of the Fibonacci sequence. It offers a glimpse into the details of Fibonacci numbers, demonstrating their link and the captivating patterns they create. Cassini's Identity, along with Binet's Formula, adds to the mathematical concept that surrounds the Fibonacci sequence, developing our awareness of this numerical concept.

3.5.3. Continued Fractions and the Convergence to the Golden Ratio

Continued fractions provide a captivating and infinite representation of numbers, and when applied to the Golden Ratio (f), they reveal a convergence. Expressing the Golden Ratio (f) as a continued fraction provides a fascinating insight into its infinite nature:

$$f = 1+\cfrac{1}{1+\cfrac{1}{1+\cfrac{1}{1+\ddots}}}]$$

This continued fraction representation reveals the self-similarity inherent in the Golden Ratio, where it can be subdivided into smaller and identical parts indefinitely.

The phenomenon of continued fractions lies in their ability to approximate numbers through partial fractions. As we extend the continued fraction representation of (f) by adding more terms, we witness a remarkable convergence toward the Golden Ratio.

Let's illustrate this convergence with a step-by-step example:

Example: Converging to the Golden Ratio (f)

1. **First Term (1):**

The initial term of the continued fraction is 1. This represents a basic approximation of (f).

2. **Second Term (1 + 1)**

Adding the second term, 1, provides a more refined approximation:

$$1+\frac{1}{1}=2$$

While not yet equal to (f), it's closer.

3. **Third Term (1 + 1 + 1):**

Continuing, the third term further refines the approximation:

$$1+\frac{1}{1+\frac{1}{1}}=\frac{3}{2}\approx 1.5$$

4. **Fourth Term (1 + 1 + 1 + 1):**

Including the fourth term narrows the gap:

$$1+\frac{1}{1+\frac{1}{1+\frac{1}{1}}}=\frac{5}{3}\approx 1.6667$$

5. **Continuing the Process:**

As we add more terms, the continued fraction approximation gradually approaches the true value of (f).

This convergence to the Golden Ratio is infinite but increasingly accurate with each additional term, demonstrating the power of continued fractions as a tool for approximating real numbers.

The use of continued fractions to approximate the Golden Ratio underscores the elegance and infinite nature of both mathematical concepts. It showcases how a simple sequence of nested fractions can provide increasingly precise approximations of a mathematical constant. This convergence to (f) exemplifies the connection between continued fractions and the Golden Ratio (Stoma et al., 2008).

3.6. BEYOND FIBONACCI: GENERALIZED FIBONACCI SEQUENCES

In our exploration of number patterns and sequences, we have discussed Fibonacci. However, mathematics offers sequences beyond the classic Fibonacci numbers. In this section, we will delve into the domain of Generalized Fibonacci Sequences, studying the Lucas Numbers, Tribonacci sequence, and higher-order linear recurrence relations (Reick, 2012).

3.6.1. Lucas Numbers

Lucas Numbers, named after the French mathematician Édouard Lucas, are a close mathematical relative of the Fibonacci sequence. They follow a similar recursive structure:

$$L(0)=2, \quad L(1)=1, \quad L(n)=L(n-1)+L(n-2)$$

Much like the Fibonacci sequence, Lucas Numbers exhibit their unique set of properties and patterns. These numbers often intertwine with Fibonacci numbers, opening avenues for mathematical exploration.

3.6.2. Tribonacci and Beyond: Generalized Sequences

While Fibonacci and Lucas numbers capture our imagination, mathematicians have extended these ideas to create a multitude of generalized sequences. One prominent member of this family is the Tribonacci sequence.

The Tribonacci sequence starts with three initial values:

$$T(0)=0, \quad T(1)=0, \quad T(2)=1$$

The recurrence relation for Tribonacci is:

$$T(n)=T(n-1)+T(n-2)+T(n-3)$$

This sequence exemplifies how we can expand the concept of Fibonacci to encompass a broader spectrum of sequences, each with its unique characteristics and mathematical concepts.

3.6.3. Higher-Order Linear Recurrence Relations

To truly grasp the power and versatility of generalized Fibonacci sequences, one must study higher-order linear recurrence relations. These relations extend the principles of the Fibonacci sequence to include more than two previous terms in the recurrence.

A general higher-order linear recurrence relation takes the form:

$$a(n) = c_1 \cdot a(n-1) + c_2 \cdot a(n-2) + \ldots + c_k \cdot a(n-k)$$

where $a(n)$ represents the nth term of the sequence, and $(c_1, c_2, \ldots, c_k)$ are constants.

These higher-order relations give rise to a number of sequences, each defined by its own set of coefficients and initial values. These sequences exhibit diverse patterns, properties, and behaviors, adding depth to the world of generalized Fibonacci sequences. Generalized sequences extend the boundaries of mathematical exploration, offering novel perspectives and deeper insights into number patterns.

3.7. CHAOS AND SELF-SIMILARITY

In mathematics and natural patterns, chaos and self-similarity reveal a profound connection that gives rise to the world of fractals. This section will discuss the intricate connection between Fibonacci numbers, chaos theory, and fractals, highlighting the fascinating concept of iterated function systems (IFS).

3.7.1. Fractals and Self-Similarity in Nature

Fractals are geometric shapes that exhibit self-similarity at different scales. This self-replicating property is pervasive in nature, from the branching patterns of trees to the coastlines of continents. To comprehend this phenomenon, we must delve into the concept of self-similarity (Figure 3.14).

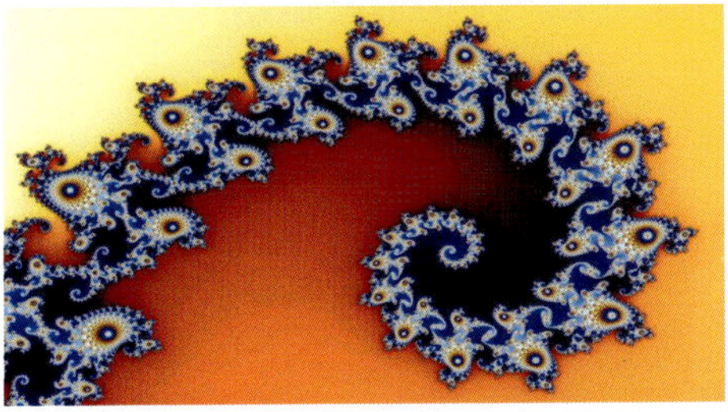

Figure 3.14. This partial view of the Mandelbrot set, possibly the world's most famous fractal.

Source: Craig Haggit, Creative Commons License.

Self-similarity is the property where a structure exhibits a repeating pattern or shape at different scales. A classic example is the Koch snowflake, a fractal curve that showcases self-similarity through its recursive construction.

3.7.2. The Connection Between Fibonacci Numbers and Chaos Theory

Fibonacci numbers, with their inherent mathematical elegance, find an unexpected link to chaos theory. Chaos theory explores the behavior of dynamic systems that are highly sensitive to initial conditions. The Fibonacci sequence itself is deterministic and predictable, but its connection to chaos lies in the Fibonacci sequence modulo (n).

Did You Know?

There's a profound connection between Fibonacci numbers and chaos theory. Explore how seemingly random and chaotic processes can have underlying order, and how this relates to the Fibonacci sequence.

When the Fibonacci sequence is taken modulo (n), it exhibits periodic behavior and chaotic patterns, depending on the value of (n). This unexpected connection between a well-behaved sequence and chaos theory underscores the intricate and surprising nature of mathematics.

3.7.3. Iterated Function Systems and the Fibonacci Sequence

IFS offer a profound link between the Fibonacci sequence and fractals. IFS is a mathematical framework used to generate fractals through a set of functions and probabilities. These functions are applied iteratively to points in space, creating intricate and often self-similar structures.

The Fibonacci sequence can be employed within an IFS to create stunning fractal patterns. Each element of the sequence can determine the probability of choosing a particular function in the IFS. By controlling these probabilities, one can craft fractals with Fibonacci-inspired structures, where self-similarity emerges through the recursive application of functions (Figure 3.15).

Figure 3.15. Example of an iterated function system in the black spleenwort fern.

Source: Larry Bradley, Creative Commons License.

The exploration of chaos and self-similarity unveils a captivating point of mathematical connections. From the self-similar fractals in nature to the unexpected relationship between Fibonacci numbers and chaos theory, mathematics once again proves its ubiquity in the natural world.

IFS add another layer of complexity, allowing us to craft fractals inspired by the Fibonacci sequence. This section has delved into the connection between these concepts, offering a glimpse into the intricate concepts of mathematics and nature.

It reminds us that the seemingly chaotic and self-similar patterns we encounter are, at their core, governed by mathematical patterns and order, enriching our understanding of the world around us.

3.8. ROBOTICS AND BIOMETRICS

From robotics to biometrics, the Fibonacci sequence and Golden Ratio find utility in unexpected places, shaping the way we design robots, recognize individuals, and analyze human behaviors. By unraveling the mathematical essence of these sequences and their real-world implications, we uncover the remarkable and ever-evolving role they play in modern science, technology, and mathematics.

3.8.1. Robotics

In robotics, these sequences find their application in the following areas:

- **Path Planning:** In robotics, the Fibonacci sequence can be used to create spiral or circular paths. For example, to calculate the coordinates (x, y) of points along a Fibonacci spiral, you can use the following equations:

$$x_n = F(n) \cdot \cos(\theta)$$
$$y_n = F(n) \cdot \sin(\theta)$$

Here, $F(n)$ represents the nth Fibonacci number, and (θ) determines the angle of rotation as you move along the spiral.

- **Manipulator Design:** The Golden Ratio (f) can be applied to optimize the proportions of robot arms and links. For instance, you can use it to determine the ideal lengths of segments in a robotic arm:

$$(L_1 = f \cdot L_2)$$

where (L_1) and (L_2) are the lengths of two consecutive segments in the robot arm.

- **Swarm Robotics:** The Fibonacci sequence can help determine the positions of robots within a swarm. For instance, you can use it to space robots evenly in a circular formation. To calculate the angular position of each robot, you can use:

$$\theta_n = \frac{2\pi}{F(n)}$$

This equation assigns each robot a unique angle based on the nth Fibonacci number.

3.8.2. Biometrics

In biometrics, these sequences find their application in the following areas:

- **Facial Recognition:** The Golden Ratio can be applied to analyze facial features. One common approach is to calculate the ratio of distances between specific features, such as the ratio of the distance between the eyes to the width of the nose. If this ratio is close to (f), it suggests a harmonious facial proportion.

$$\frac{\text{Distance between eyes}}{\text{Width of nose}} \approx f$$

- **Iris Scanning:** Fibonacci patterns can be used in iris recognition. Iris texture can be quantified using a mathematical model based on the Fibonacci sequence.

$F(n)$ can represent the number of iris textures, and $F(n+1)$ can represent the number of features within each texture.

- **Voice Analysis:** Fibonacci-based algorithms can be used to analyze speech rhythm. For instance, the duration of speech segments can be measured using Fibonacci numbers:

If $F(n)$ represents the length of a speech segment, $F(n+1)$ can represent the duration of the next segment.

- **Vascular Pattern Recognition:** Fibonacci patterns can be applied to analyze vascular structures in the hand. This can be done by quantifying the distances between vascular branching points using Fibonacci numbers.
- **Keystroke Dynamics:** Fibonacci sequences can be applied to analyze typing patterns. For instance, the time interval between keystrokes can be measured using Fibonacci numbers:

$F(n)$ can represent the time between key presses and $F(n+1)$ can represent the time interval for the next key press.

In these mathematical applications, the Fibonacci sequence and the Golden Ratio are used to quantify and analyze various aspects of robotic motion, human characteristics, and behavior for practical purposes in robotics and biometrics.

Activity 3.1

Imagine you are an architect tasked with designing a new building. You want to incorporate the aesthetic appeal of the Golden Ratio into the building's proportions. How would you use your knowledge of the Golden Ratio and Fibonacci numbers to create a visually appealing and harmonious architectural design?

3.9. SUMMARY

This chapter delves into the fascinating world of Fibonacci numbers and the mysterious Golden Ratio (f). It starts by introducing the well-known

Fibonacci sequence, which is defined by simple rules and is represented mathematically. The chapter explores how one can calculate Fibonacci numbers and explains the interesting patterns and properties inherent in this sequence. Of particular interest is the approximation of the Golden Ratio through Fibonacci ratios.

The Golden Ratio, denoted as (f), takes center stage as its mathematical expression and its relationship with consecutive Fibonacci numbers are expounded upon. Geometric constructions involving the Golden Ratio are unveiled, shedding light on its significance in various fields. Practical applications of Fibonacci numbers and the Golden Ratio emerge from their presence in nature to their influence in art, architecture, financial markets, and computer science.

The chapter then delves into deeper mathematical relationships, unveiling Binet's Formula and Cassini's Identity, which provide alternative ways to compute Fibonacci numbers. Additionally, it explores the convergence of continued fractions to the Golden Ratio. Beyond the classic Fibonacci sequence, generalized Fibonacci sequences like Lucas numbers and Tribonacci sequences are introduced, broadening the understanding of these numeric phenomena. Nature's intricate designs, embodied in the Fibonacci spiral, are discussed, showcasing the role of spirals in natural patterns such as seashells, galaxies, and hurricanes. Chaos theory enters the scene as it reveals the connection between Fibonacci numbers and chaotic systems, emphasizing the presence of order in apparent randomness. The chapter concludes by reflecting on the profound implications of Fibonacci and the Golden Ratio in describing natural order, leaving readers with philosophical and scientific questions to ponder.

REFERENCES

1. Akhtaruzzaman, M., & Shafie, A. A., (2011). Geometrical substantiation of Phi, the golden ratio and the baroque of nature, architecture, design, and engineering. *International Journal of Arts, 1*(1), 1–22.
2. Alpago, H., (2022). Analysis of the golden ratio theory in terms of economic theories. *Iğdır University Journal of Social Sciences, 1*(31), 406–418.
3. Atela, Golé, & Hotton, (2003). A dynamical system for plant pattern formation: A rigorous analysis. *Journal of Nonlinear Science, 12*, 641–676.
4. basak, R., (2022). Golden ratio and Fibonacci sequence: Universal footprints of the golden flow. *Turkish Online Journal of Design Art and Communication, 12*(4), 1092–1107.
5. Benavoli, A., Chisci, L., & Farina, A., (2009). Fibonacci sequence, golden section, Kalman filter and optimal control. *Signal Processing, 89*(8), 1483–1488.
6. Bergeron, F., & Reutenauer, C., (2017). *Golden Ratio and Phyllotaxis, What is the Mathematical Link?* arXiv preprint arXiv:1704.02880.
7. Bergeron, F., & Reutenauer, C., (2019). Golden ratio and phyllotaxis, a clear mathematical link. *Journal of Mathematical Biology, 78*, 1–19.
8. Boman, B. M., Dinh, T., Decker, K., Emerick, B., Raymond, C., & Schleiniger, G., (2017). Why do Fibonacci numbers appear in patterns of growth in nature. A model for tissue renewal based on asymmetric cell division. *Fibonacci Q, 55*(5), 30–41.
9. Boman, B. M., Ye, Y., Decker, K., Raymond, C., & Schleiniger, G., (2019). Geometric branching patterns based on p-Fibonacci sequences: Self-similarity across different degrees of branching and multiple dimensions. *Fibonacci Quart, 52019*(7), 29.
10. Ching, W. K., Cong, Y., & Tsing, N. K., (2009). A max-min principle for phyllotactic patterns. In: *Complex Sciences: First International Conference, Complex 2009, Shanghai, China, February 23–25, 2009, Revised Papers, Part 2* (Vol. 1, pp. 1329–1336). Springer Berlin Heidelberg.
11. De Bartolo, D., De Luca, M., Antonucci, G., Schuster, S., Morone, G., Paolucci, S., & Iosa, M., (2022). The golden ratio as an ecological affordance leading to aesthetic attractiveness. *PsyCh Journal, 11*(5), 729–740.

12. Debnath, L., (2011). A short history of the Fibonacci and golden numbers with their applications. *International Journal of Mathematical Education in Science and Technology, 42*(3), 337–367.
13. Dorfman, A. A., (1986). *A Theory of Form and Proportion in Music (Golden Section, Tension/Release, Symmetry, Dynamic Process, Chopin, Preludes)* (Vol. 1, pp. 1–24). University of California, Los Angeles.
14. Drăgoi, A. L., (2017) A *Simple Recursive Geometrical Construct Method and Some Properties and Occurrences of the Golden Ratio (Phi) in Mathematics, Physics, Chemistry, and Biology (Including Human Medicine), Also Related to the Transcendentals PI and to Euler's Constant (e)* (Vol. 10, pp. 1–13). Research Gate preprint. DOI.
15. Eirgash, M. A., & Toğan, V., (2023). A novel oppositional teaching learning strategy based on the golden ratio to solve the time-cost-environmental impact trade-off optimization problems. *Expert Systems with Applications, 224*, 119995.
16. Falbo, C., (2005). The golden ratio—A contrary viewpoint. *The College Mathematics Journal, 36*(2), 123–134.
17. Fanthorpe, L., Lionel, F., & Fanthorpe, P., (2013). *Mysteries and Secrets of Numerology* (Vol. 16, pp. 1–17). Dundurn.
18. Fauvel, J., Flood, R., & Wilson, R. J., (2006). *Music and Mathematics: From Pythagoras to Fractals* (Vol. 1, pp. 1–18). Oxford University Press, USA.
19. Folkerts, M., (2004). *Leonardo Fibonacci'a Knowledge of Euclid's "Elements" and of Other Mathematical Texts, 1*, 1000–1021.
20. Ghose, M., & Bahadur, B., (2019). 14 Fibonacci sequence. *Asymmetry in Plants: Biology of Handedness, 1*, 193.
21. Glushkov, S., (1976). On approximation methods of Leonardo Fibonacci. *Historia Mathematica, 3*(3), 291–296.
22. Goldstine, S., (2006). Dancing elves and a flower's view of Euclid's algorithm. *The Mathematical Intelligencer, 28*(4), 23–30.
23. Herrmann, B., (2018). Visibility in a pure model of golden spiral phyllotaxis. *Mathematical Biosciences, 301*, 185–189.
24. Hoffman, M., (1999). *Fibonacci Proportions in Composition: An Analysis of an Original Work for Jazz Orchestra* (Vol. 1, pp. 1–24). New York University.

25. Hotton, S., Johnson, V., Wilbarger, J., Zwieniecki, K., Atela, P., Golé, C., & Dumais, J., (2006). The possible and the actual in phyllotaxis: Bridging the gap between empirical observations and iterative models. *Journal of Plant Growth Regulation, 25*, 313–323.
26. Høyrup, J., (2005). Leonardo Fibonacci and Abbaco culture. A proposal to invert the roles. *Revue D'histoire des Mathématiques, 11*(1), 23–56.
27. Hretcanu, C. E., & Craşmareanu, M., (2009). Applications of the golden ratio on Riemannian manifolds. *Turkish Journal of Mathematics, 33*(2), 179–191.
28. Iosa, M., Fusco, A., Marchetti, F., Morone, G., Caltagirone, C., Paolucci, S., & Peppe, A., (2013). The golden ratio of gait harmony: Repetitive proportions of repetitive gait phases. *BioMed Research International, 2013*.
29. Iosa, M., Morone, G., & Paolucci, S., (2018). Phi in physiology, psychology, and biomechanics: The golden ratio between myth and science. *Biosystems, 165*, 31–39.
30. Jin, Z., Janoschka, D., Deng, J., Ge, L., Dreher, P., Frank, B., & Li, J., (2021). Phyllotaxis-inspired nanosieves with multiplexed orbital angular momentum. *ELight, 1*, 1–11.
31. Kazankov, V., & Gubin, V., (2020). Psychology of ustoychivost': Numerical scale for assessing human's ustoychivost' according to the golden ratio rule. In: *E3S Web of Conferences* (Vol. 210, p. 20018). EDP Sciences.
32. Kim, J. H., Cho, T. K., & Abdullaev, S., (2008). *A Study on Visualization Methods of Phyllotaxis Pattern* (pp. 129–132). Korean Multimedia.
33. King, C., (2013). Leonardo Fibonacci. *The European Mathematical Awakening: A Journey Through the History of Mathematics from 1000 To 1800*, 33.
34. Knill, O., & Tangerman, F., (2011). Self-similarity and growth in Birkhoff sums for the golden rotation. *Nonlinearity, 24*(11), 3115.
35. Kuliš, M. Š., & Hodžić, S., (2020). The golden ratio and the Fibonacci sequence in theory and practice. In: *RSEP Conferences* (Vol. 204, pp. 1–19).
36. Levitov, L., & Letters, J., (1991). Fibonacci numbers in botany and physics: Phyllotaxis. *JETP letters, 54*, 546–550.
37. MacQueen, J., (2006). The Fibonacci series and the lambda formula. In: *Complete and Full with Numbers* (Vol. 1, pp. 279–282). Brill.

38. Martinez, C. C., Chitwood, D. H., Smith, R. S., & Sinha, N. R., (2016). Left–right leaf asymmetry in decussate and distichous phyllotactic systems. *Philosophical Transactions of the Royal Society B: Biological Sciences, 371*(1710), 20150412.
39. Mayo, S. J., (2021). *Rozin B Double Helix of Phyllotaxis: Analysis of the Geometric Model of Plant Morphogenesis BrownWalker Press Irvine 2020* (Vol. 1, pp. 1–10). In: Wiley Online Library.
40. Meicenheimer, R. D., & Zagorska-Marek, B., (1989). Consideration of the geometry of the phyllotaxic triangular unit and discontinuous phyllotactic transitions. *Journal of Theoretical Biology, 139*(3), 359–368.
41. Meisner, G. B., (2018). *The Golden Ratio: The Divine Beauty of Mathematics* (Vol. 1, pp. 1–19). Race Point Publishing.
42. Minarova, N., (2014). The Fibonacci sequence: Nature's little secret. *CRIS-Bulletin of the Centre for Research and Interdisciplinary Study, 2014*(1), 7–17.
43. Moyon, M., (2019). *Teaching Mathematics and Algorithmics with Recreational Problems: The Liber Abaci of Fibonacci* (Vol. 1, pp. 1–21). In: Oslo Metropolitan University.
44. Nematollahi, A. F., Rahiminejad, A., & Vahidi, B., (2020). A novel meta-heuristic optimization method based on golden ratio in nature. *Soft Computing, 24*, 1117–1151.
45. Newell, A. C., Shipman, P. D., & Sun, Z., (2008). Phyllotaxis: Cooperation and competition between mechanical and biochemical processes. *Journal of Theoretical Biology, 251*(3), 421–439.
46. Okabe, T., Ishida, A., & Yoshimura, J., (2019). The unified rule of phyllotaxis explaining both spiral and non-spiral arrangements. *Journal of the Royal Society Interface, 16*(151), 20180850.
47. Olsen, S., Marek-Crnjac, L., He, J. H., & El Naschie, M. S., (2020). A grand unification of the sciences, art & consciousness: Rediscovering the Pythagorean Plato's golden mean number system. *Journal of Progressive Research in Mathematics, 16*(2), 2888–2931.
48. Omotehinwa, T., & Ramon, S., (2013). Fibonacci numbers and golden ratio in mathematics and science. *International Journal of Computer and Information Technology, 2*(4), 630–638.
49. Ozturk, S., Yalta, K., & Yetkin, E., (2016). Golden ratio: A subtle regulator in our body and cardiovascular system? *International Journal of Cardiology, 223*, 143–145.

50. Pandey, A. K., Kanchan, S., & Verma, A. K., (2023). Applications of Fibonacci sequences and golden ratio. *Journal of Informatics Electrical and Electronics Engineering (JIEEE), 4*(1), 1–11.
51. Parshall, K. H., (2017). A plurality of algebras, 1200–1600: Algebraic Europe from Fibonacci to Clavius. *BSHM Bulletin: Journal of the British Society for the History of Mathematics, 32*(1), 2–16.
52. Petekkaya, E., Ulusoy, M., Bagheri, H., Şanlı, Ş., Ceylan, M. S., Dokur, M., & Karadağ, M., (2021). Evaluation of the Golden Ratio in nasal conchae for surgical anatomy. *Ear, Nose & Throat Journal, 100*(1), 57–61.
53. Pletser, V., (2017). *Fibonacci Numbers and the Golden Ratio in Biology, Physics, Astrophysics, Chemistry and Technology: A non-Exhaustive Review.* arXiv preprint arXiv:1801.01369.
54. Prokopakis, E. P., Vlastos, I. M., Picavet, V., Nolst, T. G., Thomas, R., Cingi, C., & Hellings, P. W., (2013). The golden ratio in facial symmetry. *Rhinology, 51*(1), 18–21.
55. Prusinkiewicz, P., Zhang, T., Owens, A., Cieslak, M., & Elomaa, P., (2022). *Phyllotaxis Without Symmetry: What Can we Learn from Flower Heads?* In: Oxford University Press UK.
56. Reick, C. H., (2012). Self-similarity and scaling in two models of phyllotaxis and the selection of asymptotic divergence angles. *Journal of Theoretical Biology, 313*, 181–200.
57. Ricketts, R. M., (1982). The biologic significance of the divine proportion and Fibonacci series. *American Journal of Orthodontics, 81*(5), 351–370.
58. Rizzo, A., (2023). The golden ratio theorem: A framework for interchangeability and self-similarity in complex systems. *Foundations of Physics, 1*, 1–8.
59. Rozin, B., (2020). *Double Helix of Phyllotaxis: Analysis of the Geometric Model of Plant Morphogenesis*. Brown Walker Press.
60. Sadoc, J. F., Rivier, N., & Charvolin, J., (2012). Phyllotaxis: A non-conventional crystalline solution to packing efficiency in situations with radial symmetry. *Acta Crystallographica Section A: Foundations of Crystallography, 68*(4), 470–483.
61. Sah, P. K., Raj, A. M., & Sah, A., (2020). Fibonacci sequence with golden ratio and its application. *Int. J. Math. Trends Technol, 66*(3), 28–32.

62. Salingaros, N. A., (2012). Applications of the golden mean to architecture. *Meandering Through Mathematics, 21.*
63. Salis, A., & Ninham, B. W., (2014). Models and mechanisms of Hofmeister effects in electrolyte solutions, and colloid and protein systems revisited. *Chemical Society Reviews, 43*(21), 7358–7377.
64. Schwierz, N., Horinek, D., Sivan, U., & Netz, R. R., (2016). Reversed Hofmeister series—The rule rather than the exception. *Current Opinion in Colloid & Interface Science, 23*, 10–18.
65. Scott, T. C., & Marketos, P., (2014). On the origin of the Fibonacci Sequence. *MacTutor History of Mathematics, 23.*
66. Sen, S. K., & Agarwal, R. P., (2008). Golden ratio in science, as random sequence source, its computation and beyond. *Computers & Mathematics with Applications, 56*(2), 469–498.
67. Sigalotti, L. D. G., & Mejias, A., (2006). The golden ratio in special relativity. *Chaos, Solitons & Fractals, 30*(3), 521–524.
68. Stakhov, A., & Rozin, B., (2006). *The Golden Section, Fibonacci Series and New Hyperbolic Models of Nature.* Paper presented at the Visual Mathematics.
69. Stoma, S., Lucas, M., Chopard, J., Schaedel, M., Traas, J., & Godin, C., (2008). Flux-based transport enhancement as a plausible unifying mechanism for auxin transport in meristem development. *PLoS Computational Biology, 4*(10), e1000207.
70. Tahir, Ö., (2023). Can the golden ratio numbers in biochemistry and mathematics have a common explanation with nucleotide bases? *Open Access Library Journal, 10*(1), 1–13.
71. Verrelli, C. M., Iosa, M., Roselli, P., Pisani, A., Giannini, F., & Saggio, G., (2021). Generalized finite-length Fibonacci sequences in healthy and pathological human walking: Comprehensively assessing recursivity, asymmetry, consistency, self-similarity, and variability of gaits. *Frontiers in Human Neuroscience, 15*, 649533.
72. Verrelli, C. M., Romagnoli, C., Colistra, N., Ferretti, I., Annino, G., Bonaiuto, V., & Manzi, V., (2023). Golden ratio and self-similarity in swimming: Breast-stroke and the back-stroke. *Frontiers in Human Neuroscience, 17.*
73. Verrelli, C., Romagnoli, C., Jackson, R., Ferretti, I., Annino, G., & Bonaiuto, V., (2021). Front crawl stroke in swimming: Phase durations and self-similarity. *Journal of Biomechanics, 118*, 110267.

74. Wahl, M., (2023). *A Mathematical Mystery Tour: Higher-Thinking Math Tasks (Grades 5–12)*. Taylor & Francis.
75. Watson, A. R., (2017). *The Golden Relationships: An Exploration of Fibonacci Numbers and Phi, 1, 1-74.*
76. Williams, J. K., (2018). Petiole length in *Boerhavia* (*Nyctaginaceae*) approximates the golden ratio Φ. *Journal of the Botanical Research Institute of Texas, 12*(2), 667–681.
77. Winkelmann, S., Schaeffter, T., Koehler, T., Eggers, H., & Doessel, O., (2006). An optimal radial profile order based on the golden ratio for time-resolved MRI. *IEEE Transactions on Medical Imaging, 26*(1), 68–76.
78. Yin, X., & Kitazawa, M. S., (2021). Beyond Fibonacci patterns and the golden angle: Phyllotactic variations and their cellular origin. *Journal of Plant Research, 134*(3), 369–371.
79. Zeina, M. A., & Ahmed, A., (2022). The golden ratio and its impact on architectural design. *International Design Journal, 12*(2), 77–90.
80. Zheng, Y., Yang, S., & Bhunia, S., (2015). SeMIA: Self-similarity-based IC integrity analysis. *IEEE Transactions on Computer-Aided Design of Integrated Circuits and Systems, 35*(1), 37–48.

CHAPTER 4

Prime Number Patterns

Contents

4.1. INTRODUCTION

Prime numbers, often regarded as the building blocks of natural numbers, hold a distinctive place in mathematics due to their unique properties. In this chapter, we will study prime numbers, commencing with their definition, delving into the implications of the Fundamental Theorem of Arithmetic, and unveiling the uniqueness of prime factorization (Granville, 2008).

A prime number is a natural number greater than one that possesses a remarkable characteristic: it has no positive divisors other than one and itself. In a mathematical sense, a prime number (p) can be expressed as follows:

p is prime if and only if, for all positive integers (a) and (b),

if $(p=a*b)$, then either $(a=1) and (b=p) or (a=p) and (b=1)$.

To put it more simply, a prime number is an integer greater than one that cannot be formed by multiplying two smaller positive integers (Szpiro, 2004).

The Fundamental Theorem of Arithmetic, a foundation of number theory, reveals the unique structure of prime numbers and their role in the composition of all positive integers. It asserts that every positive integer greater than one can be expressed as a unique product of prime numbers, irrespective of the order of these factors (Figure 4.1).

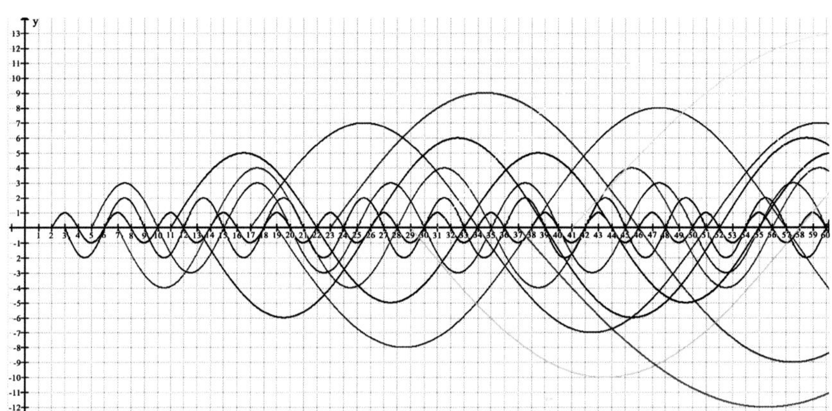

Figure 4.1. Illustration of pattern of prime numbers in the form of sinusoidal waves.

Source: Danah Henriksen, Creative Commons License.

Mathematically, the theorem can be briefly stated as follows:

"For every positive integer $(n>1)$, there exist prime numbers $(p_1, p_2, p_3, \ldots, p_k)$ (where $'k'$ can vary) and positive integers $(e_1, e_2, e_3, \ldots, e_k)$ such that:

$n = p_1^{e_1} \cdot p_2^{e_2} \cdot p_3^{e_3} \cdot \ldots \cdot p_k^{e_k}$"

This prime factorization is unique, meaning that no matter how we decompose a positive integer into prime factors, the choice of primes and the respective powers will remain the same (Cattani, 2010).

For instance, consider the number 24. Its prime factorization is $(24 = 2^3 \cdot 3^1)$, which reveals that 24 can be represented as the product of two prime numbers, 2 and 3, raised to exponents (Figure 4.2).

Factors Of 24	Factor Pairs of 24	Prime Factorisation of 24
1, 2, 3, 4, 6, 8, 12 and 24	1 × 24 2 × 12 3 × 8 4 × 6	$2^3 \times 3$

Figure 4.2. Illustration of prime factorization of number 24.

Source: BYJU's, Creative Commons License.

- **The Uniqueness of Prime Factorization:** The uniqueness of prime factorization is an integral mathematical property. It guarantees that there exists only one valid way to express any positive integer as a product of prime numbers (Weber, 2012).

Consider another example: the number 36. Its prime factorization is $(36 = 2^2 \cdot 3^2)$, signifying that 36 can be broken down into prime factors as the product of 2 and 3, each raised to a specific power. There is no alternative factorization that adheres to the Fundamental Theorem of Arithmetic.

Any attempt at an alternative factorization, such as $(36 = 2^1 \cdot 3^3)$ would not hold true since it contradicts the fundamental principle that prime factorization must be unique. The significance of the uniqueness of prime factorization reverberates throughout number theory and extends its influence into practical domains such as cryptography and computer science. This mathematical principle underscores that prime numbers are the elemental constituents of all positive integers, and their distinct factorization is a foundational aspect of the integer system (Machado & Lopes, 2020) (Figure 4.3).

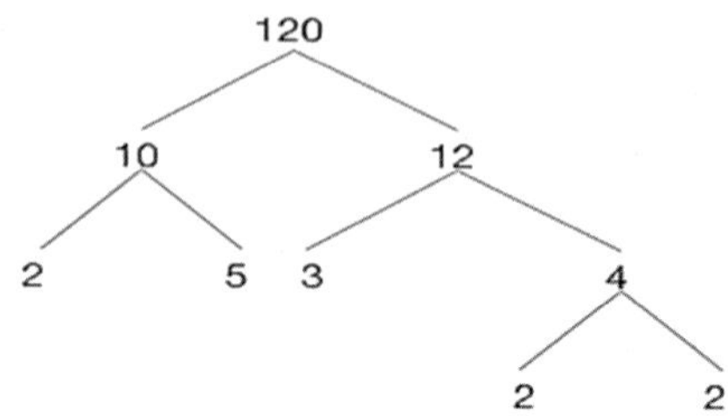

Figure 4.3. An example of prime factorization.

Source: Brilliant, Creative Commons License.

4.2. DISTRIBUTION OF PRIME NUMBERS

The distribution of prime numbers, a topic in number theory, reveals interesting patterns within mathematics. In this section, we will discuss the Prime Number Theorem, explore twin primes and prime pairs, and delve into the concept of prime gaps, including the unsolved mystery of the Bounded Gap Conjecture. These insights highlight the elusive nature of prime numbers and their seemingly unpredictable placement within the natural numbers (Peralta, 1992).

4.2.1. Prime Number Theorem

The prime number theorem is one of the most profound theorems in number theory. It offers a mathematical description of the distribution of prime numbers among the natural numbers. Formulated independently by Jacques Hadamard and Charles Jean de la Vallée-Poussin in the late 19th century, the theorem provides insights into the density and behavior of prime numbers as the natural numbers increase. In this section, we will delve into the Prime Number Theorem, its mathematical foundation, and its significance (Tao, 2011).

4.2.1.1. Asymptotic Behavior of Prime Counting Function

At the essence of the prime number theorem lies the prime counting function $p(x)$, which counts the number of prime numbers less than or equal to a given real number (x). The prime number theorem provides an asymptotic estimate for $p(x)$ as (x) grows larger. It states that:

$$\mathop{\mathrm{Lim}}_{x \to \yen} \frac{\eth(x)}{\dfrac{x}{\ln(x)}} = 1$$

In this equation, $\ln(x)$ represents the natural logarithm of (x). Essentially, the theorem tells us that as (x) becomes increasingly large, the ratio of $p(x)$ to $\left(\frac{x}{\ln(x)}\right)$ approaches 1. In other words, the density of prime numbers among the natural numbers is approximately $\left(\frac{1}{\ln(x)}\right)$ as (x) tends to infinity (Taneja, 2017).

4.2.1.2. Significance of the Prime Number Theorem

The prime number theorem has profound implications for our understanding of prime numbers and their distribution. Here are some key aspects:

1. **The "Density" of Primes:** The theorem suggests that, on average, prime numbers become less frequent as we examine larger and larger numbers. However, this does not mean that primes become scarce; rather, their distribution follows a specific pattern governed by the natural logarithm.
2. **Twin Primes:** One consequence of the prime number theorem is its connection to twin primes—pairs of primes that have a difference of 2. Although the theorem does not directly prove the existence of infinitely many twin primes, it strongly suggests their abundance within the infinite sequence of natural numbers.
3. **Approximating Prime Counts:** The prime number theorem offers a practical tool for approximating the number of primes below a certain threshold (x). By using the formula $\eth(x) \approx \frac{x}{\ln(x)}$, mathematicians can estimate prime counts efficiently, aiding in various mathematical and computational endeavors.

4.2.1.3. Illustrating the Prime Number Theorem

To illustrate the prime number theorem, consider the following example: Let $x = 1000$; using the theorem's approximation formula, we can estimate the number of primes less than or equal to 1000.

$$ð(1000) \approx \frac{1000}{\ln(1000)}$$

Calculating this expression, we find that $ð(1000) \approx 168$. This means that there are approximately 168 prime numbers between 1 and 1000. We can verify this by listing the primes in that range: 2, 3, 5, 7, 11, 13, 17, 19, 23, 29, 31, 37, 41, 43, 47, 53, 59, 61, 67, 71, 73, 79, 83, 89, 97, 101, 103, 107, 109, 113, 127, 131, 137, 139, 149, 151, 157, 163, 167, 173, 179, 181, 191, 193, 197, 199, 211, 223, 227, 229, 233, 239, 241, 251, 257, 263, 269, 271, 277, 281, 283, 293, 307, 311, 313, 317, 331, 337, 347, 349, 353, 359, 367, 373, 379, 383, 389, 397, 401, 409, 419, 421, 431, 433, 439, 443, 449, 457, 461, 463, 467, 479, 487, 491, 499, 503, 509, 521, 523, 541, 547, 557, 563, 569, 571, 577, 587, 593, 599, 601, 607, 613, 617, 619, 631, 641, 643, 647, 653, 659, 661, 673, 677, 683, 691, 701, 709, 719, 727, 733, 739, 743, 751, 757, 761, 769, 773, 787, 797, 809, 811, 821, 823, 827, 829, 839, 853, 857, 859, 863, 877, 881, 883, 887, 907, 911, 919, 929, 937, 941, 947, 953, 967, 971, 977, 983, 991, 997.

Indeed, there are 168 prime numbers within this range, aligning with the prime number theorem's prediction (Linhart &Shamir, 2009).

The prime number theorem provides profound insights into the distribution of prime numbers among the natural numbers. It offers an asymptotic description of prime density, which becomes less frequent as numbers increase.

This theorem's significance extends to estimating prime counts and illuminating the existence of twin primes, making it an indispensable tool in number theory and mathematics (Croot et al., 2015).

4.2.2. Twin Primes and Prime Pairs

Twin primes and prime pairs stand out as interesting phenomena that have long captivated mathematicians. These concepts are deeply rooted in number theory and offer unique insights into the distribution of prime numbers. Let's explore them mathematically and rigorously (Figure 4.4).

1. **Twin Primes:** Twin primes are pairs of prime numbers that are adjacent to each other, with a difference of precisely 2. In mathematical notation, $(p, p+2)$ is considered a twin prime pair if both p and p + 2 are prime numbers.

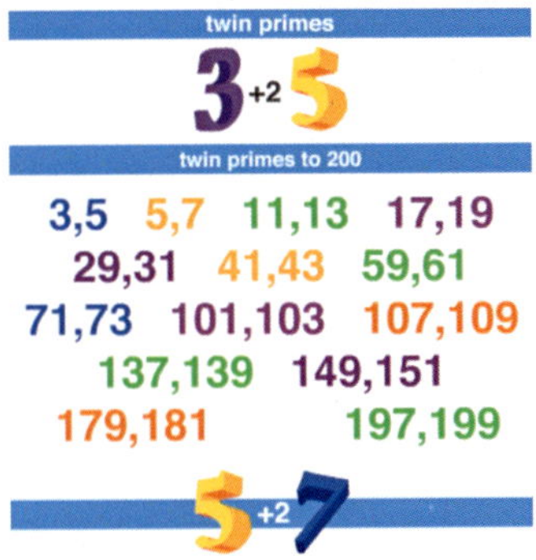

Figure 4.4. Illustration of twin prime.

Source: Math's Dictionary for kids, Creative Commons License.

2. **Prime Pairs:** Prime pairs encompass a broader category of prime number pairs. A pair (p, q) is considered a prime pair if both p and q are prime numbers, and their difference $|p-q|$ is an even number.

4.2.2.1. Infinitude of Twin Primes

The question of whether there are infinitely many twin primes is a famous unsolved problem in number theory. This conjecture suggests that there are an infinite number of twin prime pairs. Mathematically, it can be expressed as:

$$\lim_{n \to \infty} \frac{\text{number of twin primes less than or equal to n}}{\text{number of primes less than or equal to n}} = 0$$

This equation conveys the idea that as we consider larger and larger primes, the proportion of twin primes to all primes becomes vanishingly small. However, the conjecture suggests that this proportion never actually reaches zero, implying the existence of infinitely many twin primes (Pritchard et al., 1995).

4.2.2.2. Patterns and Examples

While patterns in the distribution of twin primes have been observed, proving them rigorously remains a challenge. Mathematicians have identified numerous twin prime pairs in numerical exploration. For instance:

a. (3, 5), (5, 7), (11, 13) are classic examples of twin primes.
b. (17, 19), (29, 31), (41, 43) are additional twin prime pairs.

4.2.2.3. Significance and Conjectures

Twin primes and prime pairs play a significant role in number theory and have practical applications in cryptography. One conjecture of interest is the Bounded Gap Conjecture, which proposes the existence of a positive integer k such that there are infinitely many prime pairs with a difference of 2k. In cryptography, prime pairs with larger gaps are used to enhance security in encryption algorithms. The security of many encryption methods relies on the difficulty of factoring the product of two large prime numbers, and prime pairs with larger gaps make this task computationally more challenging (Tao, 2006) (Figure 4.5).

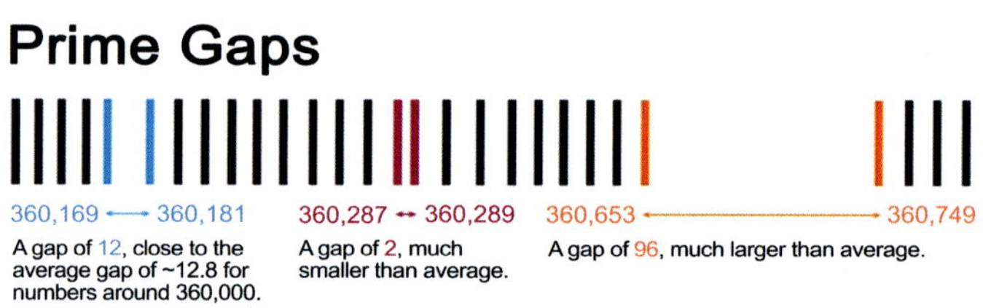

Figure 4.5. Schematic of primary gap.

Source: Wired, Creative Commons License.

4.2.2.4. Applications and Cryptography

Prime numbers, including twin primes and prime pairs, are indispensable in modern cryptography. For example, the RSA encryption algorithm relies on the difficulty of factoring the product of two large prime numbers. Prime pairs with larger gaps can be employed to enhance the security of encrypted data by increasing the computational complexity of factoring (Figure 4.6).

Figure 4.6. Image of cryptography that employs prime numbers.

Source: Joydip Kanjilal, Creative Commons License.

4.2.3. Prime Gaps and the Bounded Gap Conjecture

In the study of prime numbers, the phenomenon of prime gaps, or the spaces between consecutive prime numbers, has intrigued mathematicians for centuries. Let's delve into the mathematical intricacies of prime gaps and the Bounded Gap Conjecture (Green, 2005).

Prime gaps are the differences between consecutive prime numbers. Formally, if (p_n) represents the nth prime number, then the prime gap (g_n) is given by:

$$g_n = p_{n+1} - p_n$$

For example, if $(p_1 = 2), (p_2 = 3)$, and $(p_3 = 5)$, then the prime gaps are $(g_1 = 3 - 2 = 1)$ and $(g_2 = 5 - 3 = 2)$.

4.2.3.1. The Prime Gap Function

The prime gap function, denoted as $g(n)$, describes the prime gap at position n. It is defined as:

$$g(n) = p_{n+1} - p_n$$

This function provides insight into the distribution of prime gaps and their behavior as we examine larger and larger prime numbers (Kra, 2006).

4.2.3.2. The Bounded Gap Conjecture

The Bounded Gap Conjecture is a significant and unresolved conjecture in number theory. It proposes that there exists a positive integer k for which there are infinitely many prime pairs with a difference of 2k. Mathematically, the conjecture can be expressed as:

$$\lim_{n \to ¥} \min\{g_n, g_{n+1}\} = 2$$

In simpler terms, the conjecture suggests that, as we consider larger prime numbers, there will always be prime pairs with a gap no larger than 2k. The exact value of k remains unknown, and proving this conjecture is a formidable mathematical challenge.

4.2.3.3. Notable Prime Gaps

While prime gaps can vary widely, some are particularly noteworthy. For instance, the twin prime conjecture, a related but distinct problem, concerns prime gaps of size 2. Twin primes are prime pairs with a gap of 2, like (3, 5), (5, 7), and so on (Chang et al., 2018).

4.2.3.4. Computational Discoveries

Mathematicians have made extensive computational efforts to explore prime gaps. Advanced algorithms and high-performance computing have been used to search for large prime gaps. While there have been significant discoveries, including the verification of prime gaps with 70 million digits, proving the Bounded Gap Conjecture remains a substantial mathematical endeavor (Joux 2013).

4.2.3.5. Significance and Applications

Prime gaps and the Bounded Gap Conjecture have implications beyond number theory. They are relevant in the field of cryptography, particularly in the development of encryption algorithms that rely on the difficulty of factoring the product of two large prime numbers.

In cryptography, the security of many encryption methods, such as the RSA algorithm, depends on the challenge of factoring a large semiprime (the product of two large prime numbers). The existence of prime pairs with bounded gaps would influence the security of such encryption techniques (Lemmermeyer, 2019).

4.3. PATTERNS IN PRIME RESIDUE CLASSES

In prime numbers, exploring their distribution in residue classes offers valuable insights into their behavior. This section delves into modular arithmetic, residue classes, Euler's totient function, Fermat's little theorem, and Dirichlet's theorem on primes in arithmetic progressions. These concepts unveil remarkable patterns in prime numbers, adding depth to our understanding of their distribution.

4.3.1. Modular Arithmetic and Residue Classes

Modular arithmetic and residue classes are essential tools in the study of prime numbers, offering profound insights into their distribution and behavior. We will study these mathematical concepts, exploring their significance, equations, and their role in revealing patterns within prime numbers Shparlinski, 2018).

4.3.1.1. Modular Arithmetic

Modular arithmetic is a specialized arithmetic system that deals with remainders. It is an essential concept in number patterns and plays a crucial role in understanding prime numbers. The key idea is to perform arithmetic operations within a fixed integer "modulo" or "modulus" (Figure 4.7).

Let's denote modular arithmetic with "mod m," where m is a positive integer. In this system:

a. Addition and subtraction are performed as usual.

b. Multiplication is performed modulo m.

c. The division is performed modulo m, considering the modular multiplicative inverse when possible.

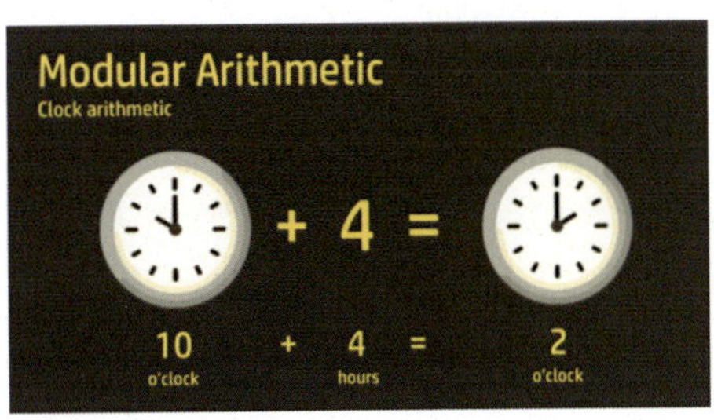

Figure 4.7. Image of modular arithmetic or clock arithmetic.

Source: Darnton, Creative Commons License.

4.3.1.2. Residue Classes

Residue classes are fundamental to modular arithmetic. They are sets of integers that are equivalent to each other when divided by a common modulus m. Formally, the residue class of an integer a modulo m is represented as [a] m and defined as:

$$[a]\text{�}=\{x \in Z : x \equiv a(mod\, m)\}$$

Here, ≡ denotes congruence, indicating that the integers in the residue class [a]m all have the same remainder when divided by m.

- **Example: Modular Arithmetic**

Let's illustrate modular arithmetic with an example. Consider performing arithmetic modulo 5. In this case:

$$8+4 \equiv 12 \equiv 2(mod\, 5)$$

$$9-6 \equiv 3 \equiv 3(mod\, 5)$$

$$3 \times 7 \equiv 21 \equiv 1(mod\, 5)$$

Modular arithmetic becomes particularly powerful when applied to prime numbers, enabling us to explore residue classes and prime patterns (Shamir, 1992).

4.3.2. Euler's Totient Function and Fermat's Little Theorem

In prime number patterns, two powerful tools—Euler's totient function and Fermat's Little Theorem—emerge as indispensable assets.

These mathematical concepts, laden with equations and precise definitions, enable us to understand patterns within prime numbers, providing crucial insights into their behavior.

4.3.2.1. Euler's Totient Function (φ)

Euler's totient function, denoted as φ (n), is a fundamental concept in number pattern and number theory. Its purpose is to quantify the number of positive integers less than or equal to n that are coprime with n. Coprime integers share no common factors other than 1. This function plays a pivotal role in understanding prime numbers and their distribution (Figure 4.8).

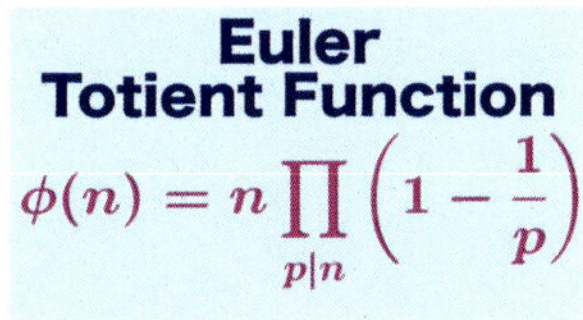

Figure 4.8. Schematic of Euler's totient function.

Source: Mohamed Omar, Creative Commons License.

The Euler's totient function φ(n) is defined as follows:

$$j\ (n) = n\left(1-\frac{1}{p_1}\right)\left(1-\frac{1}{p_2}\right)\ldots\left(1-\frac{1}{p_k}\right)$$

Where:

a. n is a positive integer.

b. $(p_1, p_2, \ldots, p_k)$ are the distinct prime factors of n.

Calculating φ (12): To illustrate Euler's totient function, let's calculate φ (12). First, we find the prime factorization of 12, which is $2^2 \times 3$. Using the formula:

$$j\ (12) = 12\left(1-\frac{1}{2}\right)\left(1-\frac{1}{3}\right) = 12\left(\frac{1}{2}\right)\left(\frac{2}{3}\right) = 4$$

So, φ (12) = 4, meaning there are four positive integers less than or equal to 12 that are coprime with 12: 1, 5, 7, and 11.

4.3.2.2. Fermat's Little Theorem

Fermat's Little Theorem is another significant theorem in a number pattern, closely related to Euler's totient function. It establishes a profound relationship between prime numbers and modular arithmetic, offering a powerful tool for primality testing (Fine & Friedman, 1976) (Figure 4.9).

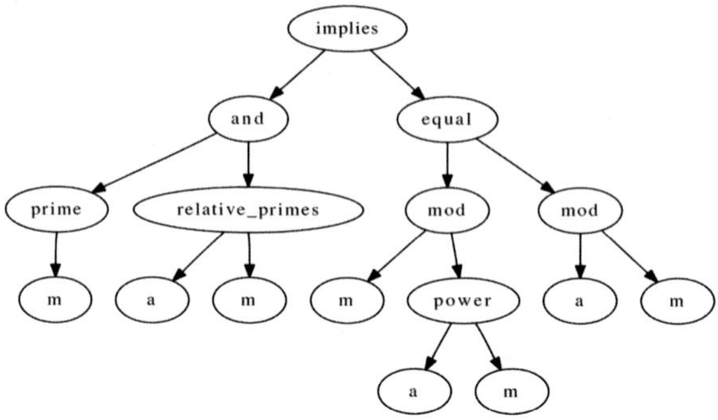

Figure 4.9. Example of the formula for Fermat's Little Theorem: prime(m) and relative primes (a, m) implies (power(a, m) mod m = a mod m).

Source: Herman Geuvers, Creative Commons License.

Fermat's Little Theorem states that if p is a prime number and a is any integer not divisible by p, then:

$$a^{p-1} \equiv 1(mod\ p)$$

In this equation:

a. "≡" denotes congruence, indicating that both sides of the equation are equivalent modulo p.

b. p is a prime number.

c. a is any integer not divisible by p.

- **Example: Applying Fermat's Little Theorem**

Let's use Fermat's Little Theorem to test the primality of 17. We will choose a = 2, and since 2 is not divisible by 17, we can apply the theorem:

$$2^{17-1} \equiv (mod\,17)$$

Now, we calculate 2^{16} modulo 17:

$$2^{16} \equiv (mod\,17)$$

Indeed, 2^{16} is congruent to 1 modulo 17. This demonstrates that 17 satisfies Fermat's Little Theorem, indicating that 17 is prime.

Euler's totient function and Fermat's Little Theorem serve as foundations in the study of prime numbers and modular arithmetic. Their applications extend beyond number patterns and number theory, reaching into cryptography and computer science.

In cryptography, Euler's totient function is crucial for encryption algorithms such as RSA. It ensures the security of these systems by making it computationally challenging to factorize large semiprime numbers. Fermat's Little Theorem is a vital tool for testing the primality of numbers, a critical step in cryptographic protocols (Pinn & Wieczerkowski, 1998).

4.3.3. Dirichlet's Theorem on Primes in Arithmetic Progressions

Dirichlet's theorem is a profound theorem in number theory that delves into the distribution of prime numbers in arithmetic progressions. This theorem shows patterns within prime numbers that extend far beyond ordinary counting. In this explanation, we will explore Dirichlet's theorem with equations, terms, and illustrative examples (Banks et al., 2013) (Figure 4.10).

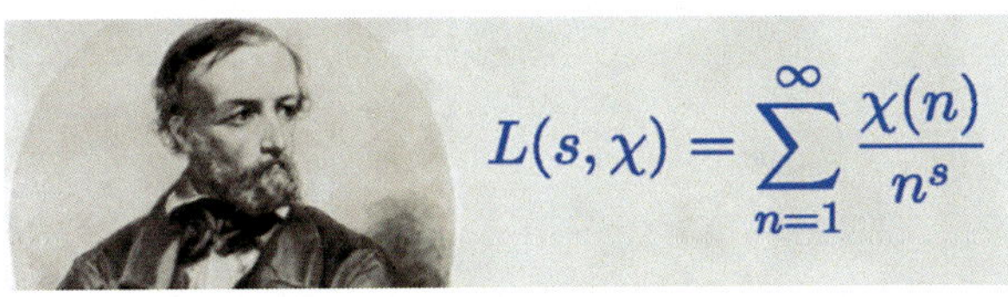

Figure 4.10. Illustration of Dirichlet's theorem and sieves.

Source: A.W. Walker, Creative Commons License.

Dirichlet's theorem, named after the German mathematician Peter Gustav Lejeune Dirichlet, addresses the existence of primes in arithmetic progressions. It states that for any two coprime positive integers a and m (i.e., integers with no common factors other than 1), there are infinitely many prime numbers that can be expressed in the form:

$$p = an + b$$

Where:

a. p is a prime number.
b. a and m are coprime positive integers.
c. n is a non-negative integer (0, 1, 2, 3,...).
d. b is an integer such that $0 \leq b < m$.

In essence, Dirichlet's theorem asserts that there are infinitely many prime numbers that are part of a specific arithmetic progression defined by a and m (Oliveira et al., 2014).

- **The General Formula:** To grasp the full scope of Dirichlet's theorem, we can express it more formally using Dirichlet L-functions. The density of primes in an arithmetic progression is given by:

$$\frac{1}{\phi(m)} \cdot \frac{1}{\log(\chi(a))}$$

Here, φ (m) is Euler's totient function, which counts the positive integers less than m that are coprime with m, and χ(a) is a character associated with the arithmetic progression defined by a and m (Maynard, 2015).

- **Example: Dirichlet's Theorem:** Let's illustrate Dirichlet's theorem with a specific example. Consider the arithmetic progression defined by a = 2 and m = 3. These numbers are coprime, as their greatest common divisor is 1. According to Dirichlet's theorem, there are infinitely many prime numbers that can be expressed in the form 2n + 1, where n is a non-negative integer.

Indeed, this holds true. The prime numbers 3, 5, 7, 11, 13, 17, and so on are all part of this arithmetic progression. As n takes on different non-negative integer values, it generates an infinite sequence of prime numbers.

Dirichlet's theorem asserts that prime numbers exhibit a structured distribution in certain arithmetic progressions defined by coprime integers a and m. This discovery not only shows patterns within prime numbers but also extends our understanding of their distribution (Goldston et al., 2009).

Furthermore, Dirichlet's theorem plays a pivotal role in advanced number patterns and number theory and is a foundation of analytic number theory. It has applications in various fields, including cryptography, where it informs the design of algorithms for secure communications (McGrath, 2022).

- **Applications and Beyond:** Dirichlet's theorem finds practical applications in number patterns, number theory, and related areas. It is instrumental in the study of L-functions, which have deep connections to prime numbers. These L-functions are crucial tools in the exploration of the Riemann Hypothesis and other unsolved problems in mathematics.

In conclusion, Dirichlet's theorem on primes in arithmetic progressions stands as a testament to the depth of number pattern and number theory. With its mathematical precision and far-reaching implications, it continues to captivate mathematicians and inspire new avenues of research in the quest to understand the complex behavior of prime numbers (Ho & Tsang, 2006).

4.4. PRIME GENERATING FUNCTIONS

In this section, we will explore various methods and functions used in the study of prime numbers. From ancient techniques like Eratosthenes' sieve to complex mathematical functions like the Riemann zeta function, we will uncover the mechanisms that help us identify and understand the distribution of prime numbers. Along the way, we will introduce the prime counting function $\pi(x)$ and demonstrate how these tools have practical applications in mathematics and beyond (Stanchescu, 2002).

4.4.1. Sieve Methods

The Sieve methods represent one of the oldest and most ingenious techniques for identifying prime numbers within a given range. We will discuss the inner workings of sieve methods, focusing primarily on Eratosthenes' sieve.

Sieve methods are mathematical techniques designed to systematically identify and isolate prime numbers within a specified range. Their historical significance is immense, as they paved the way for understanding the distribution of prime numbers long before modern computational methods emerged.

One of the earliest and most celebrated sieve methods is attributed to the Greek mathematician Eratosthenes. Born in the ancient city of Cyrene in 276 BCE, Eratosthenes devised a sieve that efficiently sifted out composite (non-prime) numbers, leaving only the prime numbers behind (Bellow et al., 1990).

4.4.1.1. Eratosthenes' Sieve: The Algorithm

Eratosthenes' sieve operates on a simple but highly effective algorithm, making it accessible even to those without advanced mathematical training. The algorithm can be summarized in a step-by-step process:

1. Create a list of integers from 2 to the desired upper limit, N, representing the numbers to be sieved.

2. Start with the first number in the list, 2, which is known to be prime.
3. Eliminate all multiples of 2 from the list. These are easily identified by iterating through the list and marking numbers that are divisible by 2 as non-prime.
4. Move to the next remaining number, 3, which is also prime.
5. Eliminate all multiples of 3 from the list, excluding 3 itself.
6. Continue this process, selecting the next unmarked number as the next prime and eliminating its multiples.
7. Repeat until you have iterated through the entire list or until the square of the current prime exceeds the upper limit, N. All remaining unmarked numbers are prime (Mancilla & Storer, 2012).

- **Example: Sieving for Primes**

Let's apply Eratosthenes' sieve to find all prime numbers up to 30:

1. Create a list of integers from 2 to 30.
2. Start with the first number in the list, 2, which is prime.
3. Eliminate all multiples of 2 from the list: 4, 6, 8, 10, 12, 14, 16, 18, 20, 22, 24, 26, 28, and 30.
4. Move to the next remaining number, 3, which is also prime.
5. Eliminate all multiples of 3 from the list: 6, 12, 18, 24, and 30.
6. Continue this process until all remaining numbers in the list are primes.

The result is a list of prime numbers from 2 to 30: 2, 3, 5, 7, 11, 13, 17, 19, 23, and 29.

4.4.1.2. Eratosthenes' Sieve: Mathematical Foundation

Eratosthenes' sieve operates on the principle of elimination. By iteratively marking the multiples of each prime number found, the sieve systematically eliminates non-prime numbers, leaving only the prime numbers untouched. The efficiency of this method lies in its ability to avoid redundant checking and eliminate multiples of primes efficiently (Seymour & Zaslavsky, 1984).

4.4.1.3. The Significance of Sieve Methods

While modern computational methods have largely replaced manual sieving, the principles of sieve methods continue to inspire mathematical exploration. Moreover, these methods have practical applications in fields such as cryptography, where prime numbers play a crucial role in ensuring the security of digital communications.

4.4.2. Generating Functions and Riemann Zeta Function

Generating functions and the Riemann zeta function are powerful mathematical tools that provide profound insights into the distribution and behavior of prime numbers (Loeliger, 1997).

4.4.2.1. Generating Functions: A Primer

Generating functions are a mathematical construct used to encode sequences of numbers into a single formal power series. They serve as a bridge between combinatorics, number theory, and analysis, offering a unified framework to study various mathematical structures, including sequences of integers.

A generating function is typically represented as a power series, often denoted by G(x), where x is a formal variable. Each term in the series corresponds to a coefficient that represents a specific term in the sequence. Generating functions can be algebraic, transcendental, or even integral, depending on the sequence they encode. The Riemann zeta function, denoted as ζ(s), is a complex function of a complex variable s. It plays a central role in number theory, particularly in understanding the distribution of prime numbers. The zeta function is defined for complex numbers s with real parts greater than 1 by the following infinite series:

$$z\left(s\right)=1^{(-s)}+2^{(-s)}+3^{(-s)}+4^{(-s)}+\ldots$$

This series converges absolutely when the real part of s is greater than 1.

Generating functions can encode information about the distribution of prime numbers through their coefficients. For example, the generating function for prime numbers can be expressed as:

$$P(x)=2^{(-s)}+3^{(-s)}+5^{(-s)}+7^{(-s)}+\ldots$$

This function encapsulates the reciprocals of prime numbers, where the coefficients correspond to the reciprocals of prime numbers raised to the power of s. By manipulating and analyzing this generating function, mathematicians can gain insights into prime number properties and distribution (Hadjicostis & Dominguez,2020).

- **Euler's Product Formula:**

One of the remarkable connections between the Riemann zeta function and prime numbers is Euler's product formula, which relates ζ(s) to the product of primes. The formula is given as:

$$Z(s) = \left(1-2^{(-s)}\right)^{(-1)} * \left(1-3^{(-s)}\right)^{(-1)} * \left(1-5^{(-s)}\right)^{(-1)} * \left(1-7^{(-s)}\right)^{(-1)} * \ldots$$

This formula reveals the link between the zeta function and the prime numbers. It demonstrates that the behavior of prime numbers is intricately connected to the analytical properties of the zeta function.

4.4.2.2. Prime-Counting Functions: The Role of ζ(s)

The Riemann zeta function is closely related to prime-counting functions, which quantify the distribution of prime numbers. One of the most famous prime-counting functions is the prime counting function π (x), which counts the number of prime numbers less than or equal to a given value x (Green & Tao, 2009). ζ(s) is intimately connected to π(x) through complex analysis. In particular, the relationship between the location of nontrivial zeros of ζ(s) and the distribution of prime numbers is a central topic in number theory. The Riemann Hypothesis, one of the most enduring unsolved problems in mathematics, deals with the location of these nontrivial zeros and their implications for prime numbers.

4.4.2.3. The Significance of Generating Functions and ζ(s)

Generating functions and the Riemann zeta function are invaluable tools in the study of prime numbers and number patterns. They offer a unified framework for understanding the distribution of primes, and their properties provide deep insights into the behavior of prime-counting functions (Pintz et al., 1989). Furthermore, the Riemann Hypothesis, linked to the zeta function, is a central open problem in mathematics, demonstrating the enduring relevance of these mathematical constructs. Solving this hypothesis would have profound implications not only for prime numbers but also for the broader field of number pattern and number theory (Goldwasser & Kilian, 1986).

4.4.3. Prime-Counting Functions

Prime counting function π(x) plays a pivotal role in the study of prime numbers and their distribution. Prime-counting functions are mathematical tools designed to quantify the distribution of prime numbers within a given range. These functions offer a systematic way to count the number of prime numbers up to a specified limit, highlighting the intricate patterns that govern the distribution of primes.

One of the most well-known prime-counting functions is the prime counting function, denoted as π(x). This function provides the count of prime numbers less than or equal to a real number x (Pintz et al., 1988) (Figure 4.11).

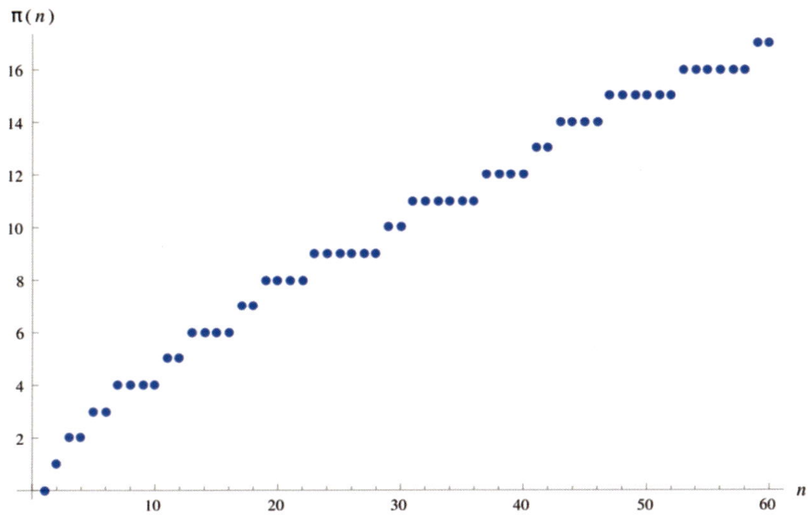

Figure 4.11. Graphical representation of prime counting function.

Source: Bender2k14, Creative Commons License.

4.4.3.1. Mathematical Expression

The prime counting function π(x) is a fundamental tool in number pattern and number theory, and its definition is straightforward: π(x) counts the number of prime numbers less than or equal to the real number x. Mathematically, it is expressed as:

$\pi(x) =$ the number of primes $\leq x$

- **Example 1: Calculating π(x)**

Let's illustrate the concept of π(x) with a simple example. To find the number of prime numbers less than or equal to 30, we can calculate π(30):

$\pi(30) =$ the number of primes ≤ 30

By listing the prime numbers less than or equal to 30 (2, 3, 5, 7, 11, 13, 17, 19, 23, and 29), we count a total of 10 primes. Therefore, π(30) = 10 (Vsemirnov, 1995).

4.4.3.2. Analytical Expressions for π(x)

While π(x) provides a clear way to count prime numbers, finding analytical expressions for it is challenging. One of the earliest attempts to approximate π(x) is the prime number theorem, which states that:

$$\left(\eth(x) \approx \frac{x}{\ln(x)} \right)$$

This approximation is remarkably accurate for large values of x, highlighting the connection between π(x) and the natural logarithm (Shevelev, 2009).

4.4.3.3. The Role of π(x) in Prime Number Theory

The prime counting function π(x) is linked to the distribution of prime numbers. Understanding its behavior and properties is crucial for addressing fundamental questions in number theory, such as the distribution of prime gaps and the existence of large prime numbers.

The relationship between the prime counting function and the Riemann zeta function ζ(s) is a central topic in number theory. The prime number theorem, mentioned earlier, relates π(x) to ζ(s) through complex analysis. Specifically, the prime number theorem states that:

$$\eth(x) \approx \int_2^x \frac{1}{\ln(t)} \, dt$$

This integral is connected to the nontrivial zeros of the Riemann zeta function, contributing to our understanding of the distribution of primes.

4.4.3.4. Prime-Counting Functions Beyond π(x)

While π(x) is the most well-known prime-counting function, it is not the only one. There exist various other functions that count primes based on specific criteria or constraints. For instance, the Chebyshev function ψ(x) counts primes with consideration for their density (Tao & Ziegler, 2023).

4.4.3.5. The Riemann Hypothesis and π(x)

The Riemann Hypothesis, one of the most prominent unsolved problems in mathematics, is closely related to π(x) and the Riemann zeta function. It suggests that all nontrivial zeros of ζ(s) have a real part equal to 1/2. Solving this hypothesis would have far-reaching implications for prime numbers and π(x).

4.4.3.6. Significance of Prime-Counting Functions

Prime-counting functions, particularly π(x), hold a central place in the field of number theory. They offer a quantitative approach to understanding prime number distribution, enabling mathematicians to explore the profound patterns and mysteries of prime numbers. In conclusion, prime-counting functions, exemplified by π(x), serve as essential tools in the study of prime numbers. Through precise mathematical definitions and analytical approximations, these functions allow us to quantify and analyze the distribution of prime numbers, contributing to our deeper understanding of one of mathematics' most interesting subjects (Gramkow, 2001).

4.5. GOLDBACH'S CONJECTURE AND PRIME PAIRS

In this section, we will study Goldbach's Conjecture, a long-standing problem in number pattern and number theory. We will explore its statement, historical context, and significance, as well as the current state of progress and research regarding this enduring conjecture.

4.5.1. Goldbach's Conjecture

Goldbach's Conjecture, one of the oldest and notable unsolved problems in number theory, poses a deceptively simple question about prime numbers. This conjecture, originally proposed by the German mathematician Christian Goldbach in a letter to Euler in 1742, revolves around the representation of even integers as sums of two prime numbers. In this exploration, we will delve into the statement of Goldbach's Conjecture, employing mathematical language, equations, and illustrative examples to shed light on the essence of this enduring puzzle.

- **The Conjecture's Statement:**

Goldbach's Conjecture, in its essence, suggests that every even integer greater than 2 can be expressed as the sum of two prime numbers. Mathematically, it can be formally stated as follows:

For every even integer n > 2, there exist prime numbers p and q such that:

$$n = p + q$$

In this statement, n represents the even integer under consideration, while p and q represent two prime numbers whose sum equals n. Essentially, the conjecture suggests that even numbers are the product of two primes in every possible way.

- **Example: Validating the Conjecture**

To illustrate the conjecture, let's examine an example: n = 28. According to Goldbach's Conjecture, 28 can be expressed as the sum of two prime numbers. Indeed, the prime numbers 5 and 23 satisfy this condition:

28 = 5 + 23

This example exemplifies the conjecture's assertion that every even integer can be decomposed into the sum of two prime numbers.

4.5.2. Progress and Current Status of Goldbach's Conjecture

Goldbach's Conjecture, a puzzle in number theory, has traversed centuries of mathematical exploration, yet it remains unproven for all, even integers.

4.5.2.1. The Historical Journey

Since its inception in the 18th century, Goldbach's Conjecture has witnessed substantial progress, albeit without a definitive resolution. Early mathematicians, including Leonhard Euler and Carl Friedrich Gauss, grappled with the conjecture, formulating ideas and proofs that laid the foundation for future investigations. Euler made significant contributions by proving the conjecture for a wide range of even integers (Behrend, 1946).

4.5.2.2. Euler's Theorem on Prime Sums

One notable milestone in the history of the conjecture is Euler's theorem, which was established by Euler in the 18th century. Euler's theorem provides a significant breakthrough by demonstrating that every even integer can be expressed as the sum of, at most, four prime numbers. Mathematically, Euler's theorem states:

Every even integer n > 2 can be expressed as the sum of either two or four prime numbers.

Euler's theorem refined the conjecture's scope and significantly advanced our understanding of it.

- **Example: Euler's Theorem**

Consider the even integer n = 48. According to Euler's theorem, we can express 48 as the sum of four prime numbers:

48 = 7 + 5 + 19 + 17

In this example, the prime numbers 7, 5, 19, and 17 sum to 48, illustrating the validity of Euler's theorem (Landman & Ventullo, 2010).

4.5.2.3. Computational Verification

The advent of computers in the 20th century ushered in a new era of exploration for Goldbach's Conjecture. Through extensive computational searches and verifications, mathematicians have confirmed the conjecture's validity for an extensive range of even integers. These computational efforts have provided substantial evidence for the conjecture's truth.

4.5.2.4. The Prime Gap Conundrum

One intriguing aspect linked to Goldbach's Conjecture is the study of prime gaps. While the conjecture asserts that every even integer can be expressed as the sum of two prime numbers, it inherently addresses the issue of prime gaps (Le, 2012).

Prime gaps have captivated mathematicians, leading to questions about the existence of large prime gaps and the overall distribution of prime numbers. The existence of arbitrarily large prime gaps has been proven, demonstrating that there are sequences of consecutive composite numbers of arbitrary length.

- **Example: The Twin Prime Conjecture**

The Twin Prime Conjecture, closely related to prime gaps, suggests that there are infinitely many pairs of twin primes (consecutive prime numbers differing by 2). While this conjecture is still unproven, it reflects the intricate nature of prime numbers and their distribution (Chen, 2016).

4.5.2.5. Current Research and Challenges

Despite centuries of exploration and the computational verification of the conjecture for vast ranges of even integers, a rigorous proof or disproof of Goldbach's Conjecture remains elusive. Current research efforts focus on refining existing techniques, exploring connections with other areas of mathematics, and investigating specific cases to inch closer to a resolution.

- **Example: Recent Progress**

In 2013, the Peruvian mathematician Harald Helfgott made significant progress by proving Goldbach's Weak Conjecture. This weaker form of the conjecture asserts that every odd integer greater than 5 can be expressed as the sum of three prime numbers. Helfgott's achievement marked a noteworthy advancement in the study of Goldbach-like problems (Conlon et al., 2014).

4.5.2.6. The Birch and Swinnerton-Dyer Conjecture

Goldbach's Conjecture is also linked to broader questions in number pattern and number theory, such as the Birch and Swinnerton-Dyer Conjecture, which concerns the analytic properties of elliptic curves. The conjecture suggests a connection between the distribution of prime numbers and the properties of certain mathematical objects.

Goldbach's Conjecture continues to intrigue mathematicians and remains a focal point of research in number pattern and number theory. While progress has been made, the conjecture's status as an unsolved problem serves as a testament to the obscurities that prime numbers hold. (Selberg, 1950).

4.6. PRIME PATTERNS IN CRYPTOGRAPHY

In this section, we explore the pivotal role of prime numbers in cryptography. We delve into their significance in RSA encryption, the essential primality testing algorithms such as Miller-Rabin, and the critical link between cryptographic security and prime factorization.

4.6.1. The Role of Prime Numbers in RSA Encryption

In modern cryptography, the RSA encryption algorithm stands as an exemplar of information security, relying fundamentally on the properties of prime numbers.

4.6.1.1. Introduction to RSA Encryption

Before delving into the role of prime numbers, let's briefly outline the RSA encryption algorithm. RSA, which stands for Rivest-Shamir-Adleman, is a widely used asymmetric encryption method. It employs a pair of keys, namely a public key and a private key, for secure data transmission and encryption. The security of RSA encryption hinges on the computational complexity of factoring the product of two large prime numbers, a task believed to be infeasible within reasonable timeframes (Romik, 2008).

4.6.1.2. The Key Generation Process

The following steps are involved in the key generation process (Figure 4.12):

1. **Prime Number Selection:** The foundation of RSA encryption lies in the selection of two distinct prime numbers, often denoted as p and q. These primes are chosen with great care, ensuring their large size and relative primality. The product n = p * q forms the modulus of both the public and private keys (p and q) are kept secret while n is made public.
2. **Public Key Computation:** The public key consists of two components – the modulus n and an encryption exponent e. The encryption exponent e is typically a small, fixed value like 65537.
3. **Private Key Computation:** The private key comprises the modulus n and a decryption exponent d. The decryption exponent d is computed using the modular multiplicative inverse of e modulo φ (n), where φ (n) represents Euler's totient function of n. Mathematically, it can be expressed as: $d \equiv e^{(-1)}\left(modj\ (n)\right)$.

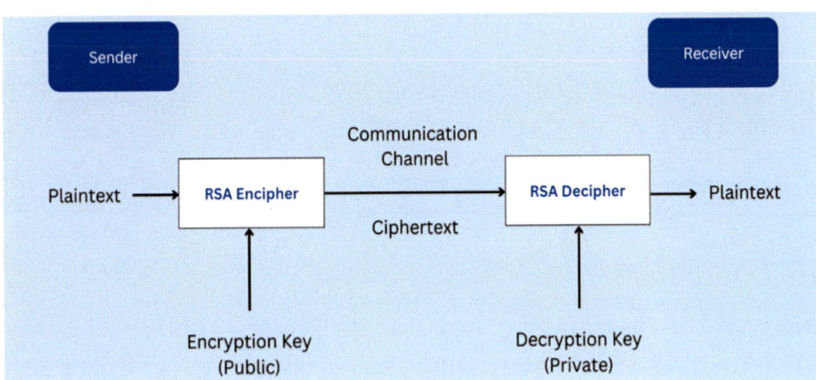

Figure 4.12. Illustration of RSA algorithm.

Source: Board infinity, Creative Commons License.

4.6.1.3. The Security Foundation: Prime Factorization

Prime factorization forms the basis of RSA's security, as breaking the encryption essentially necessitates finding p and q from n (Heule et al., 2016).

- **Example: RSA Key Generation**

Let's illustrate the process with a simplified example:

1. Select two prime numbers: p = 61 and q = 53.
2. Calculate the modulus: n = p × q = 61 × 53 = 3233.
3. Compute Euler's totient function: φ(n) = (p – 1) × (q – 1) = (61 – 1) × (53 – 1) = 60 × 52 = 3120.
4. Choose an encryption exponent: e = 17.
5. Calculate the decryption exponent using the modular multiplicative inverse of e modulo φ(n): $d \equiv 17^{(-1)}(mod\,3120)$

4.6.1.4. Encryption and Decryption

Once the keys are generated, encryption and decryption are relatively straightforward processes. When someone wishes to send a secure message to a recipient, they use the recipient's public key (n and e) to encrypt the message. The recipient, in possession of the corresponding private key (n and d), can then decrypt the message (Lipowsky et al., 2009).

4.6.1.5. The Intractability of Factoring

The security of RSA encryption rests on the belief that factoring the modulus n into its prime factors is a computationally intensive task, especially for large values of n. While there are algorithms designed to factor integers, they become increasingly inefficient as the size of n grows. As a result, RSA encryption remains a robust method for securing digital communication.

4.6.1.6. Challenges and Advancements

Over the years, researchers have explored various algorithms and techniques to factor large integers more efficiently, posing challenges to RSA's security. Consequently, the cryptographic community continually adapts by increasing key lengths and exploring alternative encryption methods.

Hence, in the digital age, the RSA encryption algorithm stands as a foundation of information security, and its foundation lies in the mathematical properties of prime numbers. The selection and manipulation of prime numbers underpin the security of RSA, ensuring that confidential data remains protected in the vast expanse of the digital landscape. While the quest for cryptographic resilience continues, RSA encryption, with its reliance on prime numbers and factorization challenges, remains a formidable safeguard for sensitive information in the modern world (Yeng et al., 1990).

4.6.2. Primality testing algorithms (e.g., Miller-Rabin)

In the world of number theory and cryptography, determining whether a given number is prime or composite is of paramount importance. The efficiency and accuracy of this determination have far-reaching implications, especially in the context of cryptographic systems that rely on prime numbers. Among the arsenal of primality testing algorithms, the Miller-Rabin primality test shines as a probabilistic method capable of swiftly identifying composites while occasionally certifying primes.

4.6.2.1. Challenges of Prime Testing

For small numbers, primality testing is straightforward and can be accomplished through trial division. However, with large numbers, traditional methods become impractical due to the sheer number of potential divisors. Hence, efficient algorithms that can swiftly ascertain primality are indispensable.

4.6.2.2. Probabilistic Primality Testing

Miller-Rabin is a probabilistic primality test, meaning that it provides a probable result. It can quickly identify most composites and correctly identify primes with high probability. The fundamental idea behind Miller-Rabin is based on Fermat's Little Theorem, which states that if p is a prime number and a is an integer not divisible by p, then:

$$a^{(p-1)} \equiv 1 \pmod{p}$$

This theorem suggests that if we choose a random integer a and find that $a^{(p-1)}$ is not congruent to 1 modulo p, then p is definitely composite. However, if $a^{(p-1)} \equiv 1 \pmod{p}$, p might be prime, or it might be a composite known as a Carmichael number.

4.6.2.3. The Miller-Rabin Algorithm

The Miller-Rabin primality test extends the idea of Fermat's Little Theorem to a more comprehensive algorithm. It checks whether the given number n is prime or composite using a series of tests based on random integers.

1. **Choose a Random Witness:** Select a random integer a from the range [2, n- 2].
2. **Compute Powers:** Calculate $a^{(n-1)}$ mod(n). If the result is not 1, n is composite, and the algorithm terminates. Otherwise, proceed to the next step.
3. **Repeated Squaring:** Square the result of the previous step k times, where k is the largest power of 2 that divides n-1. Check whether the result is congruent to 1 modulo n.

a. If it is congruent to 1, proceed to the next value of a (returning to step 1).

b. If it is not congruent to 1, and at some point, a result is not -1 modulo n, then n is composite.

4. **Repeat the Test:** Repeat steps 1–3 for a specified number of random witnesses (commonly referred to as iterations). If n passes the test for all witnesses, it is considered a probable prime.

- **Example: Applying Miller-Rabin**

Let's illustrate the Miller-Rabin algorithm with a simple example. Suppose we want to test the primality of n = 221.

1. Choose a random witness a. Let's say a = 7.
2. Calculate $7^{(220)}(mod221)$. If it's not congruent to 1, n is composite. In this case, it is congruent to 1, so we proceed.
3. Compute $7^{(110)}(mod221), 7^{(55)}(mod221)$, and finally, $7^{(27)}(mod221)$. At each step, we check for congruence to 1. If at any point it's not congruent to 1 and not -1, n is composite. In this case, all results are congruent to 1, so we proceed.
4. Repeat the test for additional random witnesses if necessary.

4.6.2.4. Accuracy and Complexity

The accuracy of the Miller-Rabin test depends on the number of random witnesses tested. With each witness, the probability of a composite number being mistakenly identified as prime decreases exponentially. Hence, multiple iterations enhance accuracy (Agarwal, 2020).

In terms of complexity, the Miller-Rabin test is relatively efficient. Its time complexity is $O(klog^3(n))$, where k is the number of iterations.

Hence, the Miller-Rabin primality test serves as a valuable tool in the domain of number theory and cryptography. While it provides probabilistic results, its ability to efficiently identify composites and often correctly certify primes makes it a practical choice for various applications. Understanding the mathematical principles behind this test is crucial for grasping its significance in the broader context of prime number theory and its pivotal role in ensuring the security of digital communication and data encryption (Sobamowo, 2016).

4.6.3. Cryptographic Security and Prime Factorization

In modern cryptography, prime factorization stands as both a foundation of security and a formidable computational challenge. The security of numerous cryptographic systems relies on the assumption that factoring large composite numbers into their prime constituents is a daunting task (Bacher, 2022).

4.6.3.1. The Basis of Cryptographic Security

The security of many cryptographic protocols, including the widely used RSA encryption, relies on the difficulty of factoring the product of two large prime numbers. The core idea is rooted in the fundamental theorem of arithmetic, which states that every positive integer greater than 1 can be uniquely represented as a product of prime numbers. For instance, the number 15 can be factored as:

$$15 = 3 * 5$$

This unique prime factorization property forms the bedrock of RSA encryption.

4.6.3.2. Prime Factorization Complexity

The complexity of prime factorization is central to the security of RSA and other cryptographic systems.

As of the time of writing this book, no known efficient algorithm can factor large composite numbers into their prime constituents in polynomial time. The best-known algorithm for general-purpose integer factorization is the General Number Field Sieve (GNFS), which has a sub-exponential running time. Its complexity is approximately $exp\left(\left(64/9\right)^{(1/3)} * \left(log\left(n\right)\right)^{(1/3)}\right)$, where n is the number to be factored.

For small numbers, factorization is straightforward. However, as numbers grow larger, the computational effort required increases exponentially. This exponential growth in complexity forms the basis of cryptographic security.

4.6.3.3. Security Parameters in Cryptography

In cryptographic applications like RSA, security parameters are chosen to ensure that the factorization of n is infeasible within a reasonable timeframe. This involves selecting large prime numbers p and q with hundreds of digits.

For example, a 2048-bit RSA key consists of two prime numbers: p and q, each with 1024 bits. Factoring the product n = p*q into its prime constituents is considered computationally infeasible with current technology (Carmichael, 1912).

4.6.3.4. The Quantum Threats

While classical computers face immense challenges in factoring large numbers, the advent of quantum computing introduces a new dimension to the field of cryptography. Shor's algorithm, developed by mathematician Peter Shor, is a quantum algorithm that can efficiently factor large numbers. If practical quantum computers become a reality, they could potentially break widely used encryption schemes like RSA (Carmichael, 1912). Cryptographic security, particularly in the context of RSA encryption, hinges on the computational complexity of prime factorization.

The mathematical foundation of prime numbers as building blocks for composite numbers forms the basis of many cryptographic systems. The formidable challenge of factoring large numbers into their prime constituents serves as a bulwark against unauthorized access to sensitive information and digital communications.

Understanding the intricate relationship between prime factorization and cryptographic security is paramount in understanding the mechanisms that underpin secure digital transactions, data protection, and information privacy in the digital age (Rankin, 1961).

4.7. SUMMARY

This chapter covers the distribution of prime numbers, introducing concepts such as the Prime Number Theorem, twin primes, and the fascinating notion of prime gaps, along with the Bounded Gap Conjecture. Moving forward, it examines patterns within prime residue classes, delving into modular arithmetic, Euler's Totient Function, and Dirichlet's Theorem on Primes in Arithmetic Progressions.

The chapter proceeds to discuss prime-generating functions, including sieve methods and their connection to the Riemann Zeta Function, leading to a discussion of prime-counting functions. It also explores the enigmatic Goldbach's Conjecture, providing insights into its progress and current status, with a particular focus on prime pairs. Finally, the chapter sheds light on the practical applications of prime number patterns in cryptography, highlighting their pivotal role in RSA encryption, primality testing algorithms such as Miller-Rabin, and the critical significance of prime factorization in ensuring cryptographic security. Throughout, the chapter emphasizes both the theoretical significance and the practical importance of prime numbers, leaving readers with a profound appreciation for this fundamental element of number theory.

REFERENCES

1. Adleman, L., & Huang, M., (1987). Recognizing primes in random polynomial time. In: *Proceedings of the Nineteenth Annual ACM Symposium on Theory of Computing* (Vol. 1, pp. 462–469).
2. Agarwal, R. P., (2020). Pythagorean theorem before and after Pythagoras. *Advanced Studies in Contemporary Mathematics, 30*(3), 357–389.
3. Andrews, W. S., & Sayles, H. A., (1913). Magic squares made with prime numbers to have the lowest possible summations. *The Monist, 1*(1), 623–630.
4. Aschbacher, M., (2004). The status of the classification of the finite simple groups. *Notices of the American Mathematical Society, 51*(7), 736–740.
5. Bacher, R., (2002). Determinants of matrices related to the Pascal triangle. *Journal de Théorie des Nombres de Bordeaux, 14*(1), 19–41.
6. Banks, W. D., Freiberg, T., & Turnage-Butterbaugh, C. L., (2013). *Consecutive Primes in Tuples*. arXiv preprint arXiv:1311.7003, 1-20.
7. Batchko, R. G., (2014). A prime fractal and global quasi-self-similar structure in the distribution of prime-indexed primes. arXiv preprint arXiv:1405.2900,1-20.
8. Behrend, F. A., (1946). On sets of integers which contain no three terms in arithmetical progression. *Proceedings of the National Academy of Sciences, 32*(12), 331, 332.
9. Bellow, A., Jones, R., & Rosenblatt, J., (1990). Convergence for moving averages. *Ergodic Theory and Dynamical Systems, 10*(1), 43–62.
10. Bennett, M. A., Martin, G., O'Bryant, K., & Rechnitzer, A., (2018). Explicit bounds for primes in arithmetic progressions. *Illinois Journal of Mathematics, 62*(1–4), 427–532.
11. Berend, D., & Bilu, Y., (1996). Polynomials with roots modulo every integer. *Proceedings of the American Mathematical Society, 124*(6), 1663–1671.
12. Bombieri, E., Friedlander, J. B., & Iwaniec, H., (1986). *Primes in Arithmetic Progressions to Large Moduli, 1*(1), 1–20.
13. Bombieri, E., Friedlander, J. B., & Iwaniec, H., (1989). Primes in arithmetic progressions to large moduli. III. *Journal of the American Mathematical Society, 2*(2), 215–224.

14. Carmichael, R. D., (1912). On composite numbers P which satisfy the Fermat congruence a p-1≡ 1 mod P. *The American Mathematical Monthly, 19*(2), 22–27.
15. Cattani, C., (2010). Fractal patterns in prime numbers distribution. In: *Computational Science and Its Applications–ICCSA 2010: International Conference, Fukuoka, Japan, March 23–26, 2010, Proceedings, Part II 10* (pp. 164–176).
16. Chang, M. C., D'Andrea, C., Ostafe, A., Shparlinski, I., & Sombra, M., (2018). Orbits of polynomial dynamical systems modulo primes. *Proceedings of the American Mathematical Society, 146*(5), 2015–2025.
17. Chen, G., (2016). On monochromatic sums of squares of primes. *Journal of Number Theory, 162*, 180–189.
18. Chua, L., Park, S., & Smith, G., (2015). Bounded gaps between primes in special sequences. *Proceedings of the American Mathematical Society, 143*(11), 4597–4611.
19. Cojori, F., (1922). Origin of the names arithmetical and geometrical progression and proportion. *School Science and Mathematics, 22*(8), 734–737.
20. Coltharp, F. L., (1966). Introducing the integers as ordered pairs. *School Science and Mathematics, 66*(3), 277–282.
21. Conlon, D., Fox, J., & Zhao, Y., (2014). The Green-Tao theorem: An exposition. *EMS Surveys in Mathematical Sciences, 1*(2), 249–282.
22. Creech-Galloway, C., Collins, B. C., Knight, V., & Bausch, M., (2013). Using a simultaneous prompting procedure with an iPad to teach the Pythagorean theorem to adolescents with moderate intellectual disability. *Research and Practice for Persons with Severe Disabilities, 38*(4), 222–232.
23. Croot, E., Lyall, N., & Rice, A., (2015). Polynomials and primes in generalized arithmetic progressions. *International Mathematics Research Notices, 2015*(15), 6021–6043.
24. Croux, C., & Haesbroeck, G., (1999). Influence function and efficiency of the minimum covariance determinant scatter matrix estimator. *Journal of Multivariate Analysis, 71*(2), 161–190.
25. Deshpande, D. S., Riccomini, P. J., Hughes, E. M., & Raulston, T. J., (2021). Problem solving with the Pythagorean theorem: A Think aloud analysis of secondary students with learning disabilities. *Learning Disabilities: A Contemporary Journal, 19*(1), 23–47.

26. Dubickas, A., (2014). Polynomials with multiple roots at 1. *International Journal of Number Theory, 10*(2), 391–400.
27. Eberly, W., Giesbrecht, M., & Villard, G., (2000). On computing the determinant and Smith form of an integer matrix. In: *Proceedings 41st Annual Symposium on Foundations of Computer Science* (Vol. 1, No. 1, pp. 675–685).
28. Elias, P., (1975). Universal codeword sets and representations of the integers. *IEEE Transactions on Information Theory, 21*(2), 194–203.
29. Ernvall-Hytönen, A. M., & Palojärvi, N., (2022). Explicit bound for the number of primes in arithmetic progressions assuming the generalized Riemann hypothesis. *Mathematics of Computation, 91*(335), 1317–1365.
30. Fine, N. J., & Friedman, D., (1976). Look, Ma, no primes. *Mathematics Magazine, 49*(5), 249–250.
31. Ford, K., Green, B., Konyagin, S., Maynard, J., & Tao, T., (2018). Long gaps between primes. *Journal of the American Mathematical Society, 31*(1), 65–105.
32. Goldston, D. A., Pintz, J., & Yildirim, C. Y., (2009). Primes in tuples I. *Annals of Mathematics, 1*(1), 819–862.
33. Goldwasser, S., & Kilian, J., (1986). Almost all primes can be quickly certified. In: *Proceedings of the Eighteenth Annual ACM Symposium on Theory of Computing* (pp. 316–329).
34. Gramkow, C., (2001). On averaging rotations. *Journal of Mathematical Imaging and Vision, 15*(1–2), 7–16.
35. Granville, A., (1987). Least primes in arithmetic progressions. In: *Number Theory, Proc. Internat. Number Theory Conf., Laval* (Vol. 1, No. pp. 306–321).
36. Granville, A., (2008). Prime number patterns. *The American Mathematical Monthly, 115*(4), 279–296.
37. Green, B., & Tao, T., (2009). A note on the Freiman and Balog–Szemerédi–Gowers theorems in finite fields. *Journal of the Australian Mathematical Society, 86*(1), 61–74.
38. Green, B., (2005). *Long Arithmetic Progressions of Primes*. arXiv preprint math/0508063, 1–20.
39. Grosswald, E., (1958). Some theorems concerning partitions. *Transactions of the American Mathematical Society, 89*(1), 113–128.

40. Hadjicostis, C. N., & Domínguez-García, A. D., (2020). Privacy-preserving distributed averaging via homomorphically encrypted ratio consensus. *IEEE Transactions on Automatic Control, 65*(9), 3887–3894.
41. Hajrizaj, D., (2009). New method to compute the determinant of a 3x3 matrix. *International Journal of Algebra, 3*(5), 211–219.
42. Heath-Brown, D. R., (1979). The differences between consecutive primes, III. *Journal of the London Mathematical Society, 2*(2), 177–178.
43. Heath-Brown, D. R., (1997). Almost-prime k-tuples. *Mathematika, 44*(2), 245–266.
44. Heule, M. J., Kullmann, O., & Marek, V. W., (2016). Solving and verifying the Boolean Pythagorean triples problem via cube-and-conquer. In: *International Conference on Theory and Applications of Satisfiability Testing* (Vol. 1, No. 1, pp. 228–245).
45. Hildebrand, A., & Maier, H., (1988). Gaps between prime numbers. *Proceedings of the American Mathematical Society, 104*(1), 1–9.
46. Ho, K. H., & Tsang, K. M., (2006). On almost prime k-tuples. *Journal of Number Theory, 120*(1), 33–46.
47. Hobby, D., & Silberger, D. M., (1993). Quotients of primes. *The American Mathematical Monthly, 100*(1), 50–52.
48. Hooley, C., (1975). A note on square-free numbers in arithmetic progressions. *Bulletin of the London Mathematical Society, 7*(2), 133–138.
49. Iovane, G., (2008). The distribution of prime numbers: The solution comes from dynamical processes and genetic algorithms. *Chaos, Solitons & Fractals, 37*(1), 23–42.
50. Joux, A., (2013). Faster index calculus for the medium prime case application to 1175-bit and 1425-bit finite fields. In: *Advances in Cryptology–EUROCRYPT 2013: 32nd Annual International Conference on the Theory and Applications of Cryptographic Techniques, Athens, Greece, May 26–30, 2013. Proceedings 32* (pp. 177–193).
51. Kim, J., & Stewart, C., (2015). Well-spaced integers generated by an infinite set of primes. *Proceedings of the American Mathematical Society, 143*(3), 915–923.
52. Ko, C. M., (2002). Distribution of the units digit of primes. *Chaos, Solitons & Fractals, 13*(6), 1295–1302.

53. Kra, B., (2006). The Green-Tao theorem on arithmetic progressions in the primes: An ergodic point of view. *Bulletin of the American Mathematical Society, 43*(1), 3–23.
54. Kubota, K. K., (1972). Pythagorean triples in unique factorization domains. *The American Mathematical Monthly, 79*(5), 503–505.
55. Lagarias, J. C., (1983). Sets of primes determined by systems of polynomial congruences. *Illinois Journal of Mathematics, 27*(2), 224–239.
56. Landman, B. M., & Ventullo, K., (2010). Avoiding monochromatic sequences with gaps in a fixed translation of the primes. *Utilitas Mathematica, 82*, 207.
57. Lê, T. H., (2012). Partition regularity and the primes. *Comptes Rendus Mathematique, 350*(9–10), 439–441.
58. Lemmermeyer, F., (2019). Composite values of irreducible polynomials. *Elemente der Mathematik, 74*(1), 36, 37.
59. Li, H., & Pan, H., (2008). *A Schur-Type Theorem for Primes*. arXiv preprint arXiv:0804.0840, 1-20.
60. Linhart, C., & Shamir, R., (2009). Faster pattern matching with character classes using prime number encoding. *Journal of Computer and System Sciences, 75*(3), 155–162.
61. Lipowsky, F., Rakoczy, K., Pauli, C., Drollinger-Vetter, B., Klieme, E., & Reusser, K., (2009). Quality of geometry instruction and its short-term impact on students' understanding of the Pythagorean theorem. *Learning and Instruction, 19*(6), 527–537.
62. Loeliger, H. A., (1997). Averaging bounds for lattices and linear codes. *IEEE Transactions on Information Theory, 43*(6), 1767–1773.
63. Machado, J. T., & Lopes, A. M., (2020). Multidimensional scaling and visualization of patterns in prime numbers. *Communications in Nonlinear Science and Numerical Simulation, 83*, 105128.
64. Mancilla, C., & Storer, R., (2012). A sample average approximation approach to stochastic appointment sequencing and scheduling. *IIE Transactions, 44*(8), 655–670.
65. Martin, A., (1913). Prime numbers in arithmetical progression. *School Science and Mathematics, 13*(9), 793–797.
66. Maynard, J., (2015). Small gaps between primes. *Annals of Mathematics, 1*(1), 383–413.

67. McCurley, K. S., (1986). Polynomials with no small prime values. *Proceedings of the American Mathematical Society, 97*(3), 393–395.
68. McGrath, O., (2022). A variation of the prime k-tuples conjecture with applications to quantum limits. *Mathematische Annalen, 384*(3, 4), 1343–1407.
69. Miyazaki, T., (2013). Generalizations of classical results on Jeśmanowicz□ conjecture concerning Pythagorean triples. *Journal of Number Theory, 133*(2), 583–595.
70. Mollin, R. A., & Williams, H. C., (1989). Quadratic non-residues and prime-producing polynomials. *Canadian Mathematical Bulletin, 32*(4), 474–478.
71. Mollin, R. A., (1997). Prime-producing quadratics. *The American Mathematical Monthly, 104*(6), 529–544.
72. Morley, F., (1894). Note on the congruence, where 2^{n+1} is a prime. *The Annals of Mathematics, 9*(1/6), 168–170.
73. Moser, L., (1953). On non-averaging sets of integers. *Canadian Journal of Mathematics, 5*, 245–252.
74. Oliveira, E. S. T., Herzog, S., & Pardi, S., (2014). Empirical verification of the even Goldbach conjecture and computation of prime gaps up to $4 \cdot 10^{18}$. *Mathematics of Computation, 83*(288), 2033–2060.
75. Overmars, A., Ntogramatzidis, L., & Venkatraman, S., (2019). A new approach to generate all Pythagorean triples. *AIMS Mathematics, 4*(2), 242–253.
76. Pagiling, S. L., (2019). Exploration of students' representation in solving Pythagorean theorem problems based on cognitive style. In: *Journal of Physics: Conference Series* (Vol. 1321, No. 3, p. 032004).
77. Pande, N. A., (2015). Analysis of primes less than a trillion. *International Journal of Computer Science & Engineering Technology, 6*(6), 332–341.
78. Peralta, R., (1992). On the distribution of quadratic residues and nonresidues modulo a prime number. *Mathematics of Computation, 58*(197), 433–440.
79. Picton, T. W., Linden, R. D., Hamel, G., & Maru, J. T., (1983). Aspects of averaging. In: *Seminars in Hearing* (Vol. 4, No. 4, pp. 327–340).
80. Pinn, K., & Wieczerkowski, C., (1998). Number of magic squares from parallel tempering Monte Carlo. *International Journal of Modern Physics C, 9*(04), 541–546.

81. Pintz, J., Steiger, W. L., & Szemerédi, E., (1989). Infinite sets of primes with fast primality tests and quick generation of large primes. *Mathematics of Computation, 53*(187), 399–406.

82. Pintz, J., Steiger, W., & Szemerédi, E., (1988). Two infinite sets of primes with fast primality tests. In: *Proceedings of the Twentieth Annual ACM Symposium on Theory of Computing* (Vol. 1, No.1, pp. 504–509).

83. Pollack, P., (2014). Bounded gaps between primes with a given primitive root. *Algebra & Number Theory, 8*(7), 1769–1786.

84. Pritchard, P. A., Moran, A., & Thyssen, A., (1995). Twenty-two primes in arithmetic progression. *Mathematics of Computation, 64*(211), 1337–1339.

85. Quadrat, J. P., Lasserre, J. B., & Hiriart-Urruty, J. B., (2001). Pythagoras' theorem for areas. *The American Mathematical Monthly, 108*(6), 549–551.

86. Randriam, H., (2019). Gaps between prime numbers and tensor rank of multiplication in finite fields. *Designs, Codes and Cryptography, 87*, 627–645.

87. Rankin, R. A., (1961). XXIV.—Sets of integers containing not more than a given number of terms in arithmetical progression. *Proceedings of the Royal Society of Edinburgh Section A: Mathematics, 65*(4), 332–344.

88. Rezaifar, O., & Rezaee, H., (2007). A new approach for finding the determinant of matrices. *Applied Mathematics and Computation, 188*(2), 1445–1454.

89. Rice, B., (2007). Primitive prime divisors in polynomial arithmetic dynamics. *Integers, 7*(1), A26.

90. Romik, D., (2008). The dynamics of Pythagorean triples. *Transactions of the American Mathematical Society, 360*(11), 6045–6064.

91. Sabia, J., & Tesauri, S., (2009). The least prime in certain arithmetic progressions. *The American Mathematical Monthly, 116*(7), 641–643.

92. Salem, R., & Spencer, D. C., (1942). On sets of integers which contain no three terms in arithmetical progression. *Proceedings of the National Academy of Sciences, 28*(12), 561–563.

93. Sayles, H. A., (1916). Even order magic squares with prime numbers. Their construction by the method of "pseudo-complementaries." *The Monist, 1*(1), 137–144.

94. Selberg, A., (1950). An elementary proof of the prime-number theorem for arithmetic progressions. *Canadian Journal of Mathematics, 2*, 66–78.
95. Seymour, P. D., & Zaslavsky, T., (1984). Averaging sets: A generalization of mean values and spherical designs. *Advances in Mathematics, 52*(3), 213–240.
96. Shamir, A., (1992). IP= PSPACE. *Journal of the ACM (JACM), 39*(4), 869–877.
97. Shanks, D., (1964). On maximal gaps between successive primes. *Mathematics of Computation, 18*(88), 646–651.
98. Shannon, A. G., & Horadam, A. F., (1971). A generalized Pythagorean theorem. *The Fibonacci Quarterly, 9*(3), 307–312.
99. Shevelev, V., (2009). *An Infinite set of Generators of Primes Based on the Rowland Idea and Conjectures Concerning Twin Primes*. arXiv preprint arXiv:0910.4676, 1-20.
100. Shparlinski, I. E., (2018). Multiplicative orders in orbits of polynomials over finite fields. *Glasgow Mathematical Journal, 60*(2), 487–493.
101. Sobamowo, M. G., (2016). On the extension of Sarrus' rule to matrices: Development of new method for the computation of the determinant of matrix. *International Journal of Engineering Mathematics, 1*(1), 2016.
102. Stanchescu, Y. V., (2002). Planar sets containing no three collinear points and non-averaging sets of integers. *Discrete Mathematics, 256*(1, 2), 387–395.
103. Szemerédi, E., (1975). On sets of integers containing no k elements in arithmetic progression. *Acta Arith., 27*(199–245), 2.
104. Szpiro, G. G., (2004). The gaps between the gaps: Some patterns in the prime number sequence. *Physica A: Statistical Mechanics and its Applications, 341*, 607–617.
105. Taneja, I. J., (2015). *Representations of Palindromic, Prime and Number Patterns* (Vol. 18, pp. 1–21). RGMIA Research Report Collection.
106. Taneja, I. J., (2017). *Multiple Choice Patterns in Prime Numbers-III, 20*, 1–109.
107. Tao, T., & Ziegler, T., (2015). Narrow progressions in the primes. *Analytic Number Theory: In Honor of Helmut Maier's 60th Birthday, 1*(1), 357–379.
108. Tao, T., & Ziegler, T., (2023). *Infinite Partial Sumsets in the Primes*. arXiv preprint arXiv:2301.10303, 1-20.

109. Tao, T., (2005). *The Dichotomy Between Structure and Randomness, Arithmetic Progressions, and the Primes.* arXiv preprint math/0512114, 1-20.

110. Tao, T., (2006). Arithmetic progressions and the primes. *Collectanea Mathematica, 1*(1), 37–88.

111. Tao, T., (2011). Structure and randomness in the prime numbers. In: *An Invitation to Mathematics: From Competitions to Research* (Vol. 1, No. 1, pp. 1–7).

112. Vartziotis, D., & Wipper, J., (2017). The fractal nature of an approximate prime counting function. *Fractal and Fractional, 1*(1), 10.

113. Vsemirnov, M. A., (1995). Infinite sets of primes, admitting Diophantine representations in eight variables. *Zapiski Nauchnykh Seminarov POMI, 220*, 36–48.

114. Wagstaff, S. S., (1979). Greatest of the least primes in arithmetic progressions having a given modulus. *Mathematics of Computation, 33*(147), 1073–1080.

115. Weber, H. J., (2012). *Remarkable and Reversible Prime Number Patterns.* arXiv preprint arXiv:1203.5227, 10-20.

116. Weintraub, S., (1977). Seventeen primes in arithmetic progression. *Math. Comp., 31*(1977), 1030.

117. Willerding, M. F., (1972). Figurate numbers. *School Science and Mathematics, 72*(2), 151–158.

118. Yanney, B. F., & Calderhead, J. A., (1896). New and old proofs of the Pythagorean theorem. *The American Mathematical Monthly, 3*(4), 110–113.

119. Yeng, S., Lin, T., & Lin, Y. F., (1990). The n-dimensional Pythagorean theorem. *Linear and Multilinear Algebra, 26*(1, 2), 9–13.

120. Zippel, R., (1990). Interpolating polynomials from their values. *Journal of Symbolic Computation, 9*(3), 375–403.

CHAPTER 5

Pascal's Triangle and Binomial Coefficients

Contents

5.1. INTRODUCTION

In combinatorics and algebra, there exists a mathematical object known as Pascal's Triangle. This triangular array of numbers, named after the eminent French mathematician Blaise-Pascal but known to scholars in various forms for centuries, contains several mathematical properties and insights. Its formation and properties are connected to the study of binomial coefficients, a fundamental concept in combinatorial mathematics. In this chapter, we will explore the depths of Pascal's Triangle and the significance of binomial coefficients in number patterns and applications (Hossan & Kibria, 2022) (Figure 5.1).

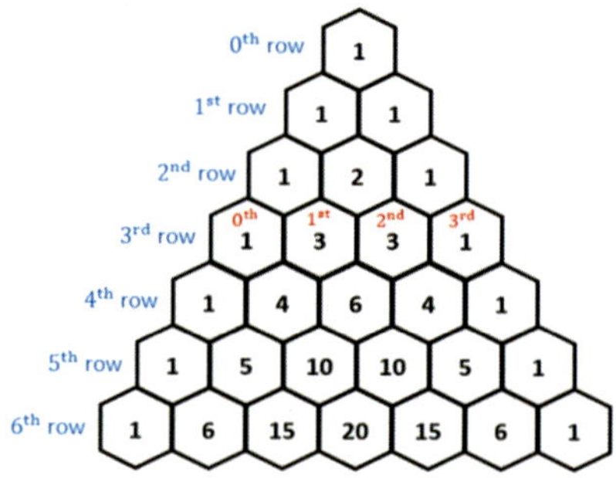

Figure 5.1. Illustration of the Pacal's triangle.

Source: Brilliant, Creative Commons License.

5.1.1. Pascal's Triangle: A Mathematical Marvel

Pascal's Triangle, an illustrative example of which is presented below for a few rows, is an arrangement of numbers where each entry is the sum of the two entries directly above it, with ones flanking the edges:

1

1 1

1 2 1

1 3 3 1

1 4 6 4 1

This simple structure conceals various mathematical concepts. To delve into this, we must first comprehend the principles underlying its construction. Each number in the triangle corresponds to a binomial coefficient, often denoted as "n choose k," which represents the number of ways to choose k items from a set of n distinct items without regard to the order of selection. This concept is central to combinatorics and probability theory and has wide-ranging applications in mathematics and beyond (Valeriu Beiu et al., 2022).

Mathematically, the binomial coefficient "n choose k" is defined as: Here, "n!" denotes the factorial of n, which is the product of all positive integers from 1 to n. The appearance of binomial coefficients in Pascal's Triangle is no coincidence; it arises from the algebraic expansion of the binomial expression , known as the binomial theorem. The coefficients of the expanded terms of this expression are precisely the entries in the nth row of Pascal's Triangle. This deep connection between binomial coefficients and algebra underscores the significance of Pascal's Triangle as a powerful mathematical tool (Fowler, 1996) (Figure 5.2).

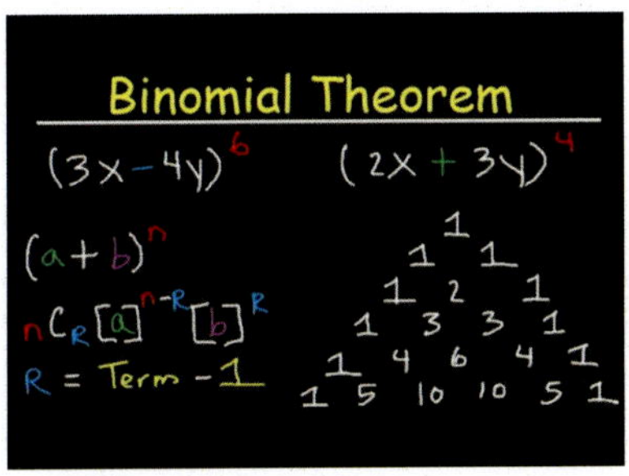

Figure 5.2. Illustration of binomial theorem.

Source: Fatima Alawneh, Creative Commons License.

5.1.2. Constructing Pascal's Triangle

Constructing Pascal's Triangle is a straightforward process that reveals its remarkable structure (Hilton & Pedersen, 1987).

1. **Row and Column Indexing:** Pascal's Triangle is constructed in a triangular grid of numbers, with rows and columns indexed starting from zero. The topmost row (Row 0) contains a single number, '1.' Each subsequent row begins and ends with '1.' To generate the numbers within the triangle, we use the following rule:
2. **Rule for Generating Pascal's Triangle:** For each number in Pascal's Triangle (except those on the edges), the value is the sum

of the two numbers directly above it in the previous row (Kuzmin & Starkov, 2021).

Mathematically, this can be expressed as:

where:

i. represents the number at row and column in Pascal's Triangle.

ii. is the number directly above and to the left in the previous row.

iii. is the number directly above in the previous row.

3. **Initializing Pascal's Triangle:** To initiate Pascal's Triangle, we start with Row 0, which contains a single number, '1.' This serves as the base (Alexey Stakhov, 2007).

4. **Constructing Subsequent Rows:** We can construct each subsequent row in Pascal's Triangle by using the Rule for Generating Pascal's Triangle, as mentioned earlier. Let's illustrate this process by constructing Row 1, Row 2, and Row 3:

i. Row 1:

a. First element (Column 0) = 1 (always starts with 1).

b. Apply the rule:

c. Last element (Column 1) = 1 (always ends with 1).

ii. Row 2:

a. First element (Column 0) = 1 (always starts with 1).

b. Apply the rule:

c. Apply the rule again:

d. Last element (Column 2) = 1 (always ends with 1).

iii. Row 3:

a. First element (Column 0) = 1 (always starts with 1).

b. Apply the rule:

c. Apply the rule again:

d. Apply the rule again:

e. Last element (Column 3) = 1 (always ends with 1).

5. **Repeating the Process:** This process can be repeated indefinitely, generating each row of Pascal's Triangle by applying the Rule for Generating Pascal's Triangle to the previous row. The resulting triangle exhibits fascinating patterns, symmetries, and mathematical properties, making it a valuable tool in various

branches of mathematics, including combinatorics, probability theory, and algebra. Its structure reveals the coefficients of the binomial expansion, among other applications, making it a fundamental concept in mathematics.

5.1.3. Patterns in Pascal's Triangle

Pascal's Triangle, a triangular array of numbers, holds several mathematical patterns and properties. (Alexey Stakhov, 2007).

5.1.3.1. Binomial Coefficients and Diagonals

One of the most fundamental patterns in Pascal's Triangle is the distribution of binomial coefficients along its diagonals. Binomial coefficients denoted as represent the number of ways to choose items from a set of distinct items without regard to order. These coefficients are crucial in combinatorics and algebra.

Each row of Pascal's Triangle corresponds to the coefficients of the binomial expansion of where is the row number. For example, Row 3 of the triangle contains the coefficients for the expansion of which is

5.1.3.2. Symmetry in Pascal's Triangle

Another pattern in Pascal's Triangle is its symmetry. If you draw a line through the apex (the number 1 at the top) vertically, you will notice that the numbers on either side of the line are mirror images of each other. This symmetry is evident in both the coefficients themselves and their placement in the triangle (Islam et al., 2022).

Mathematically, the symmetry in Pascal's Triangle can be expressed as:

This symmetry arises because choosing items from is equivalent to choosing items from

5.1.3.3. Sum of Rows

Pascal's Triangle also exhibits a simple pattern when it comes to summing the numbers in each row. The sum of the numbers in any row of Pascal's Triangle is equal to where is the row number. This can be proven by induction (Lay-Yong, 1980).

Mathematically, the sum of the numbers in Row can be represented as:

This property underscores the combinatorial significance of Pascal's Triangle, as it shows the total number of ways to choose items from a set of distinct items (Figure 5.3).

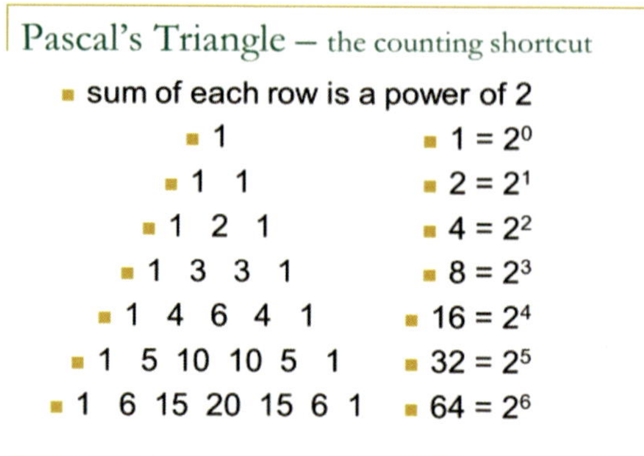

Figure 5.3. Schematic illustration of the sum of rows.

Source: Muriel Powers, Creative Commons License.

5.1.3.4. Triangular Numbers

Another pattern in Pascal's Triangle is the appearance of triangular numbers along its left edge. Triangular numbers are those that can be arranged in the shape of an equilateral triangle. They are closely related to binomial coefficients. The nth triangular number is given by and it can be observed that the numbers along the left edge of Pascal's Triangle follow this pattern (Granville, 1997).

Mathematically, the nth triangular number can be represented as:

This connection between triangular numbers and Pascal's Triangle adds a geometric dimension to the patterns within the triangle.

5.1.3.5. Fibonacci Numbers

Surprisingly, Pascal's Triangle also contains a hidden connection to the Fibonacci sequence. The Fibonacci numbers appear in Pascal's Triangle when you sum the diagonal elements starting from the second element (i.e., not counting the '1' at the apex). This sum yields the Fibonacci numbers (Skurnick & Martin, 2012) (Figure 5.4).

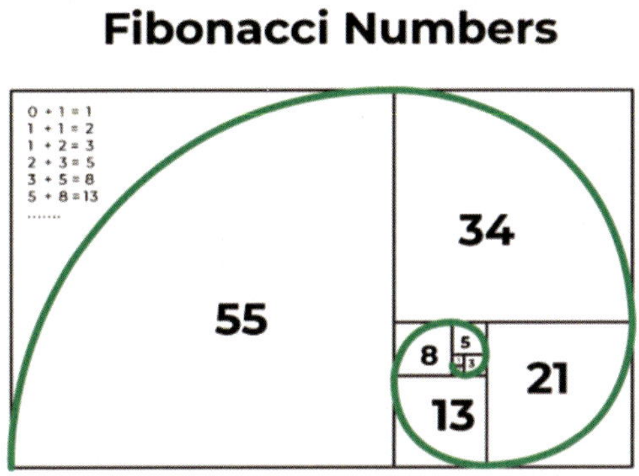

Figure 5.4. Illustration of Fibonacci numbers in Pascal's triangle.

Source: Geeks for Geeks, Creative Commons License.

Mathematically, the connection between Pascal's Triangle and the Fibonacci sequence can be represented as:

where:

i. represents the nth Fibonacci number.

ii. This unexpected link demonstrates the deep interplay of mathematical patterns.

5.1.3.6. Hockey Stick Pattern

Another interesting pattern known as the "Hockey Stick Pattern" emerges in Pascal's Triangle. This pattern resembles the shape of a hockey stick, hence its name. It is formed by selecting a number from any row and then tracing diagonally down and to the right, including all the numbers in the path. The number at the end of this diagonal is the sum of the selected number and all the numbers above it in the same column. This pattern is a manifestation of the relationship between binomial coefficients (Skurnick et al., 2012).

Mathematically, the Hockey Stick Pattern can be represented as:

where:

i. is a selected number in Pascal's Triangle.

ii. The sum on the left-hand side includes all the numbers in the diagonal path.

5.1.3.7. Fractal-Like Structure

While not immediately apparent, Pascal's Triangle exhibits fractal-like properties when zoomed in on specific regions. Each triangle within the larger triangle is itself a Pascal's Triangle, and this nesting can continue infinitely. This recursive, self-similar structure is reminiscent of fractals in mathematics (Yurkin, 2013) (Figure 5.5).

Figure 5.5. Schematic of fractal-like structure.

Source: mathtourist.blogspot, Creative Commons License.

Mathematically, this fractal-like property can be expressed as an infinite nesting of Pascal's Triangles within Pascal's Triangles:

Where:

and represent indices within the nested triangle structure.

Hence, from binomial coefficients and their combinatorial significance to symmetry, row sums, triangular numbers, Fibonacci numbers, and the intriguing Hockey Stick Pattern, this triangular arrangement of numbers continues to captivate mathematicians and serves as a source of mathematical exploration and discovery. Its patterns underpin various areas of mathematics, making it a fundamental concept in the field (Fraenkel & Kontorovich, 2007).

5.2. BINOMIAL COEFFICIENTS

Binomial coefficients play a pivotal role in various mathematical applications, including the binomial theorem, Pascal's Triangle, probability theory, and algebraic expansions. In this section, we will study the significance and properties of binomial coefficients, presenting their critical role in solving a wide array of mathematical problems and equations.

5.2.1. Definition of Binomial Coefficients

Binomial coefficients denoted as are fundamental mathematical entities that arise in the context of combinatorics and algebra (Fraenkel & Kontorovich, 2007).

Tip

When working with binomial coefficients, it's essential to understand that they represent the number of ways you can choose "k" items from a set of "n" items without regard to the order of selection. This concept is fundamental in combinatorics and probability.

5.2.1.1. Combinatorial Interpretation

At its core, a binomial coefficient represents the number of ways to choose items from a set of distinct items without regard to order. This combinatorial interpretation is fundamental to understanding the essence of binomial coefficients.

Mathematically, can be defined as:

Where:

i. represents the factorial of which is the product of all positive integers from 1 to

ii. is the factorial of which is the product of all positive integers from 1 to

iii. is the factorial of which is the product of all positive integers from 1 to

5.2.1.2. Factorial Function

The factorial function is a key component of the binomial coefficient definition. It is defined as the product of all positive integers from 1 to Mathematically:

For example,

5.2.1.3. Permutations and Combinations

To grasp the significance of binomial coefficients, it is essential to differentiate between permutations and combinations. Permutations refer to the arrangements of items in a specific order, while combinations concern the selection of items without regard to order. Binomial coefficients deal with combinations (Lovasz et al., 2003).

The formula for permutations denoted as is:

In permutations, the order of selection matters, leading to a different number of possibilities than combinations. Binomial coefficients, on the other hand, are concerned with combinations and disregard the order of selection (Clark, 1982).

5.2.1.4. The Role of and

In the definition of binomial coefficients, both and play significant roles. The value of represents the number of items to be chosen from the set of while represents the number of items not chosen. These two values together ensure that every possible combination is considered.

- **Example:**

Suppose you have a set of 5 distinct items (A, B, C, D, E), and you want to choose 3 of them. If you choose items A, B, and C, there are ways to do so. Alternatively, you could choose items D, E, and either A, B, or C. The choice of 3 items automatically implies the non-choice of 2 items, making and complementary in binomial coefficients.

5.2.1.5. Binomial Coefficients in Algebra

Binomial coefficients also play a crucial role in algebraic expressions, particularly in the expansion of binomial expressions.

The binomial theorem provides a formula for expanding expressions of the form where is a positive integer (Liu & Kuo, 2016).

The binomial theorem states: This formula expresses the expansion of a binomial expression as a sum of terms, where each term is formed by multiplying a binomial coefficient a power of (raised to and a power of (raised to . Binomial coefficients determine the coefficients of each term in the expansion (Liu & Kuo, 2016).

5.2.2. Notation and Symbolism

In mathematics, precise notation and symbolism are essential for conveying complex ideas and relationships concisely.

When dealing with binomial coefficients, a set of standardized notations and symbols are used to represent these coefficients and their properties. Understanding this notation is crucial for working with binomial coefficients effectively (Kallós, 2006).

5.2.2.1. Binomial Coefficient Notation

The most common notation for binomial coefficients is This notation represents the number of ways to choose items from a set of distinct items without regard to order. The and in are known as the arguments of the binomial coefficient, and they specify the size of the set and the number of items to be chosen, respectively (Clark, 1982) (Figure 5.6).

$$C(n,k) = \frac{n!}{k!(n - k)!}$$

Figure 5.6. Depiction of notation of binomial coefficient.

Source: Learning about Electronics, Creative Commons License.

The notation simplifies the expression of combinatorial choices and forms the foundation for many mathematical calculations involving binomial coefficients.

5.2.2.2. Alternative Notations

While is the standard notation for binomial coefficients, alternative notations are also occasionally used. These notations include:

i. This is a common shorthand notation for and is widely accepted in the mathematical literature. It represents the same concept of choosing items from items.

ii. In some contexts, especially when discussing binomial distributions in probability theory, may be used to denote binomial coefficients.

iii. Another way to represent binomial coefficients, especially in combinatorics and statistics, is

These alternative notations serve the same purpose as and are used interchangeably based on the preference of the mathematician or the context in which they are applied (Annamalai, 2022b).

5.2.2.3. Subscript and Superscript

The notation can be further clarified using subscript and superscript. The is often placed in the subscript position, while the is placed in the superscript position to provide a clear indication of their roles. For instance, represents the binomial coefficient with as the total number of items and as the number of items to be chosen (Rogers, 2006). Mathematically, is equivalent to and can be defined as:

This notation aids in understanding the roles of the arguments in the binomial coefficient expression.

5.2.2.4. Common Binomial Coefficient Values

Certain binomial coefficients have specific values that frequently appear in mathematical calculations. These values are often memorized due to their recurrence in various mathematical contexts. Some common values include:

i. The number of ways to choose 0 items or all items from a set of items is always 1.

ii. The number of ways to choose 1 item or items from a set of items is equal to

iii. The number of ways to choose 2 items from a set of items, often used in combinations and probability calculations.

Understanding these common values can simplify mathematical computations involving binomial coefficients (Song & Deng, 2021).

In conclusion, binomial coefficients are denoted using standardized notations and symbols such as and These notations are vital for accurately representing the number of ways to choose items from a set and are used in combinatorics, probability theory, algebra, and various mathematical applications. The subscript and superscript arrangement, as in clarifies the roles of the arguments, making it easier to understand their significance in the context of binomial coefficients. Additionally, recognizing common binomial coefficient values is essential for efficient mathematical calculations (Leavitt, 2011).

5.2.3. Combinatorial Interpretations

Combinatorial interpretations provide a profound understanding of the significance of binomial coefficients They connect these coefficients to counting problems and combinatorial situations, highlighting the fundamental role binomial coefficients play in enumerative combinatorics (Spresser & Fritz, 1996).

5.2.3.1. Counting Subsets and Combinations

One of the most intuitive combinatorial interpretations of involves counting subsets or combinations of a set. Suppose we have a set of distinct items, and we want to determine the number of ways to choose items from this set without regard to order. This situation is frequently encountered in combinatorics and provides the solution.

Mathematically, the combinatorial interpretation can be expressed as:

This interpretation aligns with the definition of binomial coefficients and serves as a foundation for various combinatorial problems (de Weger, 1997).

5.2.3.2. Counting Arrangements with Repetition

Another combinatorial interpretation of binomial coefficients arises when we need to count arrangements with repetition. Consider a situation where we have distinct items, and we want to arrange them in a sequence of positions, allowing repetition of items. The question becomes: How many such arrangements are possible?

Mathematically, the combinatorial interpretation for this scenario is:

The logic behind this interpretation is that we can view the problem as selecting positions from a sequence of positions for placing the distinct items. Binomial coefficients capture this count efficiently.

5.2.3.3. Counting Paths in Grids

Binomial coefficients also find application in counting paths in grids, particularly when moving only in two directions, such as up and to the right. Consider a grid with rows and columns, where we start at the bottom-left corner and want to reach the top-right corner by moving either up or to the right. The question is: How many different paths can we take?

Mathematically, the combinatorial interpretation for this grid problem is:

The reason behind this interpretation lies in the fact that we need to make moves in total, and we must choose of those moves to be "up" while the rest are "right" moves (Liu & Kuo, 2016).

5.2.3.4. Counting Multisets and Distributions

Multisets are sets that allow for the repetition of elements, and binomial coefficients are instrumental in counting multisets. Suppose we have distinct items, and we want to create multisets of size where each item may appear more than once. The question becomes: How many distinct multisets of size can we form?

Mathematically, the combinatorial interpretation for counting multisets can be represented as:

This interpretation aligns with the concept that we are essentially distributing identical items into distinct categories, where each category can receive zero or more items.

5.2.3.5. Counting Solutions to Linear Diophantine Equations

Binomial coefficients also emerge when counting the solutions to linear Diophantine equations, which are equations where integer solutions are sought. Consider a linear Diophantine equation of the form where and are integers, and we want to find the number of non-negative integer solutions for and This problem can be solved using binomial coefficients (Leavitt, 2011) (Figure 5.7).

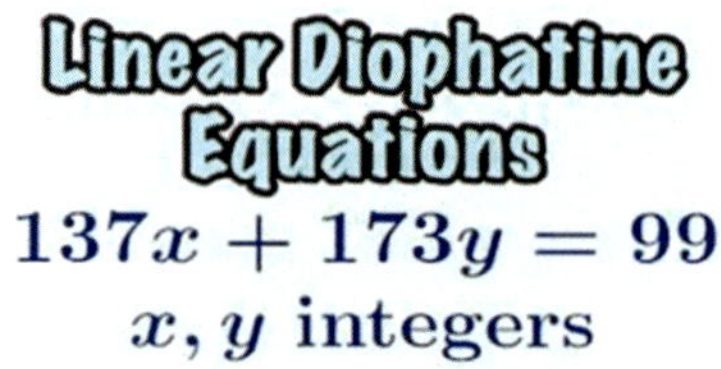

Figure 5.7. Illustrative example of linear Diophantine equation.

Source: Mohamed Omar, Creative Commons License.

Mathematically, the combinatorial interpretation for counting solutions to such equations is:

Where:

i. represents the desired sum (in the equation, corresponds to

ii. is the number of variables; in the equation, corresponds to 2 variables, and

The combinatorial interpretation arises from considering the distribution of items, similar to the multiset's interpretation.

5.2.3.6. Counting Inversions in Permutations

In combinatorics, an inversion in a permutation occurs when a larger number precedes a smaller number. Binomial coefficients come into play when counting the number of permutations with a specific number of inversions (Lu, Zhang, & Li, 2019) (Figure 5.8).

Mathematically, the combinatorial interpretation for counting inversions in permutations is:

This interpretation is particularly useful in understanding the relationships between permutations and inversion patterns.

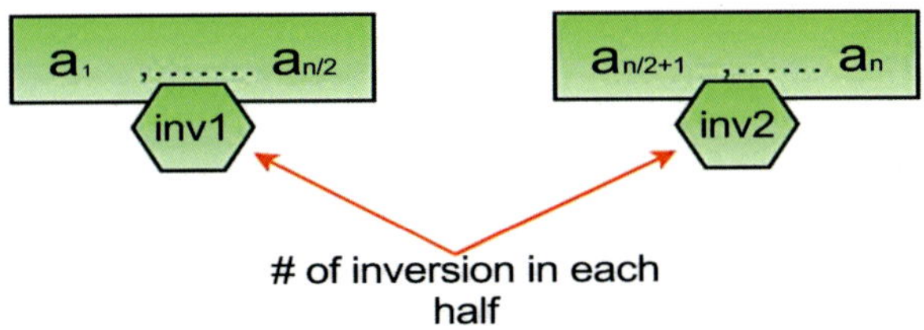

Figure 5.8. Schematic of counting of permutations.

Source: Geeks for Geeks, Creative Commons License.

5.2.3.7. Counting Distribution of Distinct Objects into Distinct Boxes

Binomial coefficients also find application in counting the distribution of distinct objects into distinct boxes, where each box may contain zero or more objects. Suppose we have distinct objects and distinct boxes. The question is: How many ways can we distribute the objects into the boxes?

Mathematically, the combinatorial interpretation for this distribution problem is (Neuwirth, 1998): The rationale behind this interpretation is like the multisets interpretation, where we are distributing objects into categories (boxes), allowing for empty categories.

5.3. PASCAL'S TRIANGLE AND BINOMIAL EXPANSION

Pascal's Triangle and binomial expansion are connected mathematical concepts. Pascal's Triangle is a triangular array of numbers where each number is the sum of the two numbers directly above it, starting with a single '1' at the apex. This triangle holds various mathematical patterns, but one of its most significant applications is in the expansion of binomial expressions. The coefficients in the expansion of known as binomial coefficients, are precisely the numbers found in Pascal's Triangle. Understanding this relationship between Pascal's Triangle and binomial expansion is crucial for algebraic manipulations and solving various mathematical problems, making these concepts fundamental in mathematics (Rushdi & Al-Amoudi, 2019).

5.3.1. Binomial Theorem

The Binomial Theorem is a fundamental mathematical principle that explains the expansion of a binomial expression, such as where is a positive integer. It reveals the algebraic expression of this binomial raised to a power, providing a formula to efficiently compute its terms. The Binomial Theorem plays a pivotal role in algebraic manipulations, combinatorics, and calculus, and it is deeply intertwined with Pascal's Triangle and binomial coefficients (Wright, 1995).

5.3.1.1. General Statement of the Binomial Theorem

The Binomial Theorem states that for any positive integer the expansion of can be expressed as (Figure 5.9):

Where:

i. represents the binomial coefficient, which is the number of ways to choose items from a set of distinct items without regard to order.

ii. and are constants.

iii. The summation symbol indicates that we sum the terms for all values of from 0 to

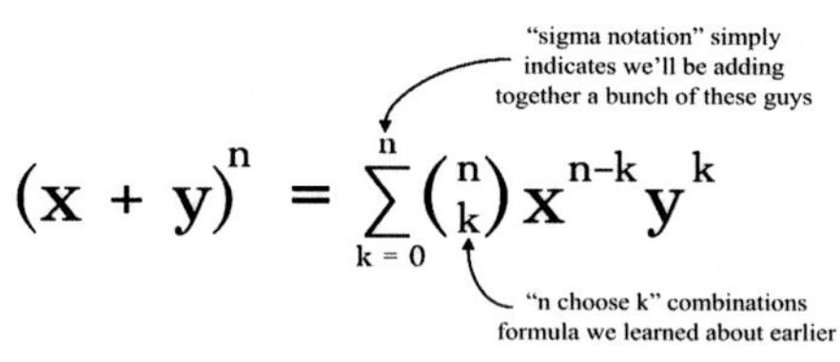

Figure 5.9. Schematic of formula of binomial theorem.

Source: Brett Berry, Creative Commons License.

5.3.1.2. Understanding the Terms

In the Binomial Theorem, the expression is expanded into a sum of terms, each of which comprises three components:

Let's break down these components (Donnelly et al., 2018):

i. This is the binomial coefficient, which represents the number of ways to choose items from a set of distinct items without regard to order (as already discussed). It is determined by the values of and and is obtained from Pascal's Triangle.

ii. This term represents the power to which (a) is raised in the expansion. It starts at (a^n) for $(k=0)$ and decreases by 1 with each successive term until it reaches $(a^0)\, for\, (k=n)$.

iii. b^k : Similarly, this term represents the power to which (b) is raised in the expansion. It starts at $(b^0)\, for\, (k=0)$ and increases by 1 with each successive term until it reaches (b^n) for $(k=n)$.

5.3.1.3. Applications of the Binomial Theorem

The Binomial Theorem finds applications in various mathematical fields (Russell, 2022):

1. **Combinatorics:** It is fundamental for counting combinations and arrangements of objects. The binomial coefficients in the expansion represent these counts.

2. **Probability Theory:** The expansion of $(a+b)^n$ provides the probabilities of various outcomes in binomial experiments.
3. **Algebra:** It simplifies algebraic expressions involving powers of binomials, making calculations more manageable.
4. **Calculus:** The Binomial Theorem is used in calculus to approximate functions, especially when dealing with small values of ( ) and (b), and to derive power series expansions.
5. **Number Theory:** The coefficients in the expansion are related to number theory and combinatorial identities.

5.3.1.4. Special Cases and Expansions

The Binomial Theorem encompasses several notable special cases and expansions:

1. **Positive Integer Exponents:** When (n) is a positive integer, the Binomial Theorem provides a finite expansion. For example, when $(n=3)$, the expansion of $(a+b)^3$ is:

$$(a+b)^3 = C(3,0)\cdot a^3 + C(3,1)\cdot a^2\cdot b + C(3,2)\cdot a\cdot b^2 + C(3,3)\cdot b^3$$

2. **Infinite Series:** The Binomial Theorem can also be applied to infinite series when \(n\) is a positive real number. For instance, when $\left(n=\frac{1}{2}\right)$, the expansion of $(1+x)^{\frac{1}{2}}$ is an infinite series known as the binomial series:

$$(1+x)^{\frac{1}{2}} = 1 + \frac{1}{2}x - \frac{1}{8}x^2 + \frac{1}{16}x^3 - \frac{5}{128}x^4 + \ldots$$

3. **Negative Exponents:** The Binomial Theorem can also be extended to negative integer exponents using the idea of binomial series.

5.3.2. Calculating Coefficients

The calculation of binomial coefficients relies on the factorial function, and it can be accomplished using the formula $C(n,k) = \frac{n!}{k!(n-k)!}$. However, it is often more practical to compute binomial coefficients using Pascal's Triangle or recursion formulas (Gwetta, Pacurar, & Smith, 2012).

- **Pascal's Triangle:** Pascal's Triangle is an invaluable tool for finding binomial coefficients.

  The triangle is constructed such that each number in the triangle is *the* sum of the two numbers directly above it. The top of the triangle is '1,' and each row begins and ends with '1.' To calculate binomial coefficients using Pascal's Triangle, locate the row corresponding to (n) (starting with the 0th row), and then select the (k)th entry in that row, where (k) ranges from 0 to (n)(Lovász et al., 2003).

For example, to find $C(5,2)$, locate the 5th row in Pascal's Triangle and count two entries from the left, resulting in $C(5,2)=10$.

Remember

In binomial expansion, Pascal's Triangle provides a handy shortcut for finding the coefficients of each term in the expanded expression. This can save you a lot of time when dealing with complex algebraic expressions.

- **Recursion Formula:** Binomial coefficients can also be computed using a recursive formula. This formula relates $C(n,k)$ to $C(n-1,k-1)$ and $C(n-1,k)$, as follows:

$$C(n,k) = C(n-1,k-1) + C(n-1,k)$$

This recursion allows for the calculation of binomial coefficients by working from smaller values of (n) and (k) towards the desired values. The initial values $C(n,0) = C(n,n) = 1$ serves as the base cases.

5.3.2.1. Symmetry in Binomial Coefficients

Binomial coefficients exhibit a remarkable symmetry, known as Pascal's Rule. This symmetry reflects the equivalence of $C(n,k)$ and $C(n,n-k)$, meaning that choosing (k) items from (n) is equivalent to choosing $(n-k)$ items from (n). Mathematically, this symmetry can be expressed as:

$$C(n,k) = C(n,n-k)$$

This symmetry is not only a key property of binomial coefficients but is also a reflection of the symmetric nature of binomial expressions.

5.3.2.2. Coefficients in Binomial Expansion

The coefficients in binomial expansion play a crucial role in expressing the expanded form of a binomial expression $(a+b)^n$. Each term in the expansion is a product of a binomial coefficient $C(n,k)$, (a) raised to a power $(n-k)$, and (b) raised to a power (k). The general term in the expansion is given by:

$$T(k) = C(n,k) \cdot a^{n-k} \cdot b^k$$

Where:

i. $T(k)$ represents the kth term in the binomial expansion.

ii. $C(n,k)$ is the binomial coefficient corresponding to the k th term.

iii. a^{n-k} represents (a) raised to the power of $(n-k)$.

iv. b^{k} represents (b) raised to the power of (k).

These coefficients determine the relative contribution of each term to the overall expression and provide a systematic way to expand and simplify binomial expressions.

5.3.2.3. Significance of Coefficients in Binomial Expansion

The coefficients in binomial expansion have several significant implications and applications:

1. **Combinatorial Significance:** Binomial coefficients count combinatorial objects such as combinations, permutations, and arrangements, making them indispensable in counting and enumerative combinatorics.
2. **Probability Theory:** In probability theory, binomial coefficients appear in the probability mass function (PMF) of the binomial distribution, which models the number of successes in a fixed number of independent Bernoulli trials.
3. **Algebraic Manipulations:** Coefficients enable the simplification of algebraic expressions involving powers of binomials, facilitating polynomial manipulations and factorizations.
4. **Calculus:** Binomial expansion is a fundamental tool in calculus for approximating functions and for deriving power series representations, which are crucial in Taylor series and Maclaurin series.
5. **Number Theory:** Binomial coefficients are related to number theory and are central to the study of combinatorial identities, partitions, and divisibility properties (Spresser & Fritz, 1996).

5.3.3. Applications in Algebra

The applications of Pascal's Triangle and binomial expansion in algebra are extensive and powerful, as they provide a framework for simplifying expressions, solving equations, and exploring the properties of polynomials.

5.3.3.1. Simplifying Binomial Expressions

One of the fundamental applications of binomial expansion in algebra is simplifying binomial expressions. Binomial expressions are algebraic expressions that consist of two terms, such as $(a+b)$. Using the Binomial Theorem, we can expand these expressions for any positive integer exponent (n). For instance, $(a+b)^3$ expands to:

$$(a+b)^3 = a^3 + 3a^2b + 3ab^2 + b^3$$

This expanded form allows us to simplify complex expressions involving binomials. For example, consider the expression $(2x-3y)^4$. We can use the expanded form to find specific terms quickly, which simplifies calculations in algebraic manipulation.

5.3.3.2. Finding Specific Terms

In algebraic equations or expressions, it is often necessary to isolate and analyze specific terms of a binomial expansion. Pascal's Triangle provides an efficient means to determine the coefficients associated with specific terms in an expansion. For example, to find the coefficient of $\left(x^2y^2\right)$ in the expansion of $(x+y)^4$, we can consult the corresponding entry in Pascal's Triangle, which is 6. This means that the coefficient of $\left(x^2y^2\right)$ in the expansion is 6. This ability to identify specific terms is valuable when solving equations or evaluating polynomial expressions (Lu, Zhang, & Li, 2019).

5.3.3.3. Algebraic Identities

Binomial expansion gives rise to several important algebraic identities that are used extensively in algebraic manipulations. One such identity is the binomial theorem itself, which states:

$$(a+b)^n = \sum_{k=0}^{n} C(n,k) \cdot a^{n-k} \cdot b^k$$

This identity provides a systematic way to expand binomial expressions of the form $(a+b)^n$ for any positive integer (n). Additionally, it leads to various binomial-related identities, such as the Vandermonde identity:

$$\sum_{k=0}^{r} C(m,k) \cdot C(n,r-k) = C(m+n,r)$$

These algebraic identities are instrumental in solving equations and simplifying expressions, contributing significantly to the field of algebra.

5.3.3.4. Polynomial Coefficients

Pascal's Triangle and binomial expansion are invaluable tools for understanding the coefficients of polynomial expressions. For example, when expanding $(a+b)^n$, the coefficients in the expansion correspond to the coefficients of the terms in the binomial expansion of (a^n) and (b^n). This knowledge helps in polynomial factorization and manipulation (Allenby & Slomson, 2010). Consider the polynomial $P(x)=(x+2)^3$. By applying the binomial expansion, we can determine its coefficients:

$$P(x)=(x+2)^3 = x^3 + 6x^2 + 12x + 8$$

This information is vital when finding roots, determining multiplicity, and analyzing the behavior of polynomial functions.

5.3.3.5. Polynomial Roots and Solutions

Binomial expansion aids in finding roots and solutions to polynomial equations. When a binomial expression is equated to zero, its solutions correspond to the roots of the polynomial equation. The ability to expand and manipulate binomial expressions is essential for solving polynomial equations efficiently (Allenby & Slomson, 2010).

For example, consider the equation $(x+3)(x-1)=0$. Expanding the left side of the equation yields $(x^2+2x-3=0)$, which is a quadratic equation. By applying algebraic techniques such as factoring or the quadratic formula, we can find the roots of this equation as $(x=-3)$ and $(x=1)$.

5.3.3.6. Binomial Theorem in Algebraic Calculations

The Binomial Theorem provides a systematic method for approximating functions and expressions involving binomials. It is especially valuable in algebraic calculations, where an exact solution is challenging to obtain.

By expanding a binomial expression, we can approximate the value of the expression for various values of the variables involved (Ghent, 1967).

For example, to estimate $(1.01)^{10}$, we can use the Binomial Theorem with $(n=10):[(1.01)^{10} \approx 1+10(0.01)=1.10$

This approximation is quite close to the actual value, demonstrating the utility of the Binomial Theorem in algebraic calculations.

5.3.3.7. Algebraic Proof and Verification

Pascal's Triangle and binomial expansion serve as tools for algebraic proof and verification of mathematical statements and identities. By utilizing the expanded forms of binomial expressions, mathematicians can establish the validity of various mathematical propositions.

For instance, consider the identity $C(n,0)+C(n,1)+C(n,2)+\ldots+C(n,n)=2^n$. To prove this identity algebraically, one can expand $(1+1)^n$ using the Binomial Theorem, which yields the sum of binomial coefficients. This algebraic approach provides rigorous proof of identity (Neuwirth, 1998).

5.4. PROPERTIES OF PASCAL'S TRIANGLE

Pascal's Triangle possesses numerous intriguing properties that are vital in various mathematical contexts. Some of the key properties include symmetry, where each row is a palindrome; the fact that each row sums to powers of 2; its relationship to binomial coefficients, which are used extensively in combinatorics; and its role in binomial expansion, where the coefficients in the expansion correspond to the entries in the triangle. Additionally, Pascal's Triangle exhibits the triangular number pattern in the diagonal elements and connects to number theory through properties such as the Fibonacci sequence. These properties make Pascal's Triangle an indispensable tool in mathematics, serving as a source of insight into combinatorics, algebra, and beyond (Uptegrove, 2015).

5.4.1. Symmetry in Pascal's Triangle

Symmetry is a prominent and fascinating property of Pascal's Triangle. This symmetry is not only esthetically pleasing but also deeply connected to the combinatorial and algebraic properties of the triangle. In this exposition, we will explore the various aspects of symmetry within Pascal's Triangle and its mathematical implications.

5.4.1.1. Symmetry about the Central Axis

One of the most apparent forms of symmetry in Pascal's Triangle is the symmetry about its central axis. If we draw a vertical line down the center of the triangle, separating it into two halves, we observe that the entries on each side of the line are mirror images of each other. This central axis of symmetry bisects the triangle, creating two symmetric halves.

Mathematically, this symmetry can be expressed as:

$$C(n,k) = C(n,n-k)$$

Where:

$C(n,k)$ represents the binomial coefficient at row (n) and position (k).

The right-hand side, $C(n,n-k)$, represents the binomial coefficient with the complementary position on the other side of the central axis.

This symmetry underscores the balance and elegance of Pascal's Triangle.

5.4.1.2. Palindromic Rows

The central axis of symmetry gives rise to palindromic rows within Pascal's Triangle. A row is considered palindromic if its entries read the same from left to right as they do from right to left. These rows are particularly notable because they exhibit a unique and elegant pattern.

For example, the third row of Pascal's Triangle is $(1,2,1)$, which is a palindromic row. Similarly, the fifth row is $(1,4,6,4,1)$, displaying the same palindromic property (Bardzell & Poimenidou, 2011).

The palindromic nature of these rows is a direct consequence of the symmetry about the central axis. It occurs because the $coefficients\, C(n,k)$ and $C(n,n-k)$ are equal, leading to symmetry not only between the left and right sides of the triangle but also within individual rows (Figure 5.10).

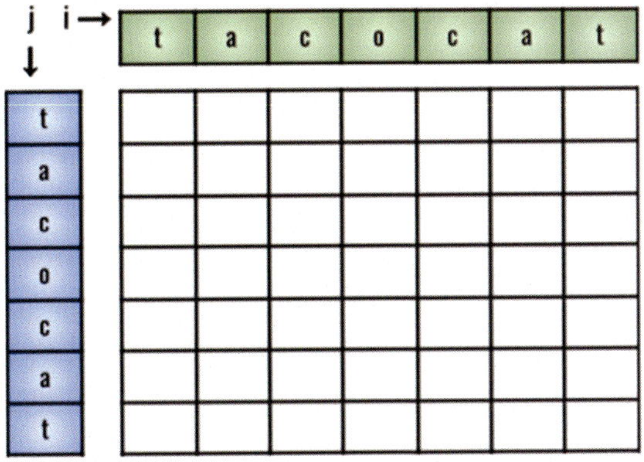

Figure 5.10. Illustration of the palindromic rows.

Source: Pranav Gautam, creative common license.

5.4.1.3. Combining Rows

Another manifestation of symmetry in Pascal's Triangle is the ability to obtain a new row by combining two symmetric rows. This operation can be performed by adding corresponding terms from two symmetric rows to generate a new row with entries equal to the sums of the corresponding entries in the original rows (Konvalina, 2000).

Mathematically, the operation can be expressed as follows:

$$C(n+1,k)=C(n,k-1)+C(n,k)$$

Where:

i. $C(n,k)$ represents an entry in row (n) at position (k).

ii. $C(n+1,k)$ represents the corresponding entry in the next row $(n+1)$ at position (k).

This operation is a consequence of the symmetry between adjacent rows in Pascal's Triangle. It allows us to generate new rows efficiently by combining existing ones, making the construction of the triangle highly systematic.

5.4.1.4. The Relationship to Binomial Coefficients

The symmetry in Pascal's Triangle is intricately connected to binomial coefficients $C(n,k)$, which are the numbers found within the triangle. Binomial coefficients represent the number of ways to choose (k) items from a set of (n) distinct items without regard to order. The symmetry within the triangle arises because $C(n,k)$ is equal to $C(n,n-k)$.

Mathematically, this relationship can be expressed as:

$$C(n,k)=C(n,n-k)$$

This relationship is fundamental in understanding and utilizing binomial coefficients and is a direct consequence of the symmetry of Pascal's Triangle. It underscores the idea that selecting (k) items is equivalent to selecting $(n-k)$ items from a set of (n) distinct items.

5.4.2. Differences in Pascal's Triangle

Pascal's Triangle, a remarkable triangular array of numbers, exhibits a multitude of mathematical properties, including patterns involving sums and differences of its elements. These properties are central to understanding the relationships between the numbers within the triangle and their applications in mathematics. In this explanation, we will explore the sums and differences within Pascal's Triangle and elucidate their mathematical significance (Cobeli & Zaharescu, 2013).

5.4.2.1. Differences between Adjacent Numbers

Examining the differences between adjacent numbers in Pascal's Triangle reveals yet another interesting pattern. If we subtract each number from the one immediately to its right in a row, we obtain a new row that forms a sequence of consecutive natural numbers, starting from 1. This sequence is known as the arithmetic sequence or the counting of numbers (Gould, 1969).

Mathematically, the differences between adjacent numbers in a row can be expressed as:

$$C(n,k)-C(n,k-1)=k$$

Where:

i. $C(n,k)$) represents the binomial coefficient at row (n) and position (k).

ii. k represents the consecutive natural numbers obtained when subtracting adjacent numbers.

For example, consider the sixth row of Pascal's Triangle, which is $(1,5,10,10,5,1)$. The differences between adjacent numbers in this row are $(5-1=4),(10-5=5),(10-10=0),(5-10=-5)$, and $(1-5=-4)$, respectively, forming the arithmetic sequence $(4,5,0,-5,-4)$.

This pattern of differences between adjacent numbers is noteworthy for its regularity and connection to the natural numbers, providing insight into the underlying structure of Pascal's Triangle (Rushdi & Al-Amoudi, 2019).

5.4.2.2. Combining Rows

The differences between adjacent numbers within Pascal's Triangle can also be used to generate new rows by combining existing rows. This operation involves adding corresponding differences in adjacent positions to create a new row with differences that form a sequence of consecutive natural numbers.

Mathematically, the operation can be expressed as follows:

$$C(n,k)-C(n,k-1)+C(n,k+1)-C(n,k)=1$$

Where:

i. $C(n,k)$ represents the binomial coefficient at row (n) and position (k).

ii. The expression on the left side represents the differences between adjacent numbers at positions $(k) and (k+1)$.

iii. The result on the right side is always 1, indicating the regularity of this operation.

For example, consider the seventh row of Pascal's Triangle, which is $(1,6,15,20,15,6,1)$. The differences between adjacent numbers are $6-1=5, 15-6=9, 20-15=5, 15-20=-5, 6-15=-9$, and $1-6=-5$. Adding the differences in adjacent positions results in $5+9=14, 9+5=14, 5+(-5)=0, (-5)+(-9)=-14$, and $(-9)+(-5)=-14$. These differences create a new row with the sequence $(14,14,0,-14,-14)$, which illustrates the pattern of consecutive natural numbers (Wright, 1995).

5.4.3. Diagonal Sums in Pascal's Triangle

Pascal's Triangle exhibits a distinct set of patterns related to diagonal sums. These sums, derived by adding the numbers along specific diagonals of the triangle, reveal interesting mathematical relationships and connections. In this exposition, we will discuss the concept of diagonal sums within Pascal's Triangle and explain their mathematical significance.

5.4.3.1. Primary Diagonal Sums

The most fundamental diagonal in Pascal's Triangle is the primary diagonal, which consists of the entries where the row number equals the position number (both starting from 0). The sums along the primary diagonal make a sequence of natural numbers, starting from 1 and increasing by 1 with each subsequent row. Mathematically, the sum along the primary diagonal can be expressed as:

$$\sum_{k=0}^{n} C(n,k) = 2^n$$

Where:

i. $C(n,k)$ represents the binomial coefficient at row (n) and position (k).

ii. The summation symbol (Σ) indicates that we sum all the numbers along the primary diagonal from $(k=0)$ to (n).

For example, in the third row of Pascal's Triangle, which is $(1,2,1)$, the sum along the primary diagonal is $(1+2+1=4)$, which corresponds to (2^2). In the fourth row, the sum along the primary diagonal is $(1+3+3+1=8)$, corresponding to (2^3).

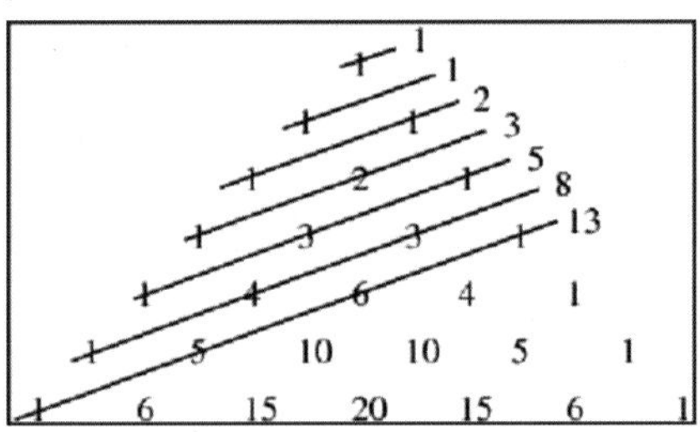

Figure 5.11. Schematic of the diagonal sum of Pascal's triangle.

Source: Klee Irwin, Creative Commons License.

The primary diagonal sums have a profound combinatorial significance. They represent the total number of ways to choose items from a set, making them a central concept in combinatorics. This relationship between diagonal sums and powers of 2 underscores the bridge between discrete mathematics and arithmetic within Pascal's Triangle (Lovász et al., 2003) (Figure 5.11).

5.4.3.2. Secondary Diagonal Sums

Moving beyond the primary diagonal, Pascal's Triangle also reveals interesting patterns in the sums along the secondary diagonals. These secondary diagonals run parallel to the primary diagonal, but each is one step to the right of the previous diagonal. The sums along the secondary diagonals form a sequence of triangular numbers, where each term is the sum of consecutive natural numbers, starting from 1 and increasing by 1 with each subsequent diagonal. Mathematically, the sum along a secondary diagonal can be expressed as:

$$\sum_{k=0}^{n} C(n,k) = \binom{n+1}{2}$$

Where:

i. $C(n\ k)$ represents the binomial coefficient at row (n) and position (k).

ii. The expression $\binom{n+1}{2}$ represents a binomial coefficient and, therefore, the sum along a secondary diagonal.

For example, in the third row of Pascal's Triangle, the sum along the first secondary diagonal is $(1+2=3)$, which corresponds to $\binom{3+1}{2}=\binom{4}{2}=6,$ a triangular number. Similarly, in the fourth row, the sum along the first secondary diagonal is $(1+3+3=7)$, which corresponds to $\binom{4+1}{2}=\binom{5}{2}=10,$ another triangular number.

The secondary diagonal sums exhibit a fascinating connection to triangular numbers, which are a common topic in number theory and combinatorics. These sums also reveal the natural progression of triangular numbers within Pascal's Triangle (Smith, 1973).

5.4.3.3. Additional Diagonal Sums

Beyond the primary and secondary diagonals, Pascal's Triangle continues to unveil patterns in diagonal sums. Each diagonal sum follows a unique mathematical sequence, and the sums themselves hold specific combinatorial interpretations.

For instance, the sums along diagonals formed by skipping every second number, such as the third diagonal $(1, 3, 6, 10, \ldots)$, correspond to tetrahedral numbers, which represent the number of ways to choose items from a set when repetitions are allowed.

Furthermore, the sums along diagonals formed by skipping every third number, such as the fourth diagonal $(1, 4, 10, 20, \ldots)$, correspond to pentatope numbers, which are related to hyperdimensional geometries and combinatorial arrangements.

These additional diagonal sums exemplify the diverse range of mathematical patterns and combinatorial interpretations that emerge from Pascal's Triangle. They extend beyond triangular numbers and tetrahedral numbers, revealing the versatility of the triangle in various mathematical contexts.

5.5. ADVANCED TOPICS

In this section, we will discuss some advanced topics related to Pascal's Triangle, exploring the profound mathematical concepts that emerge from this triangular array of numbers.

These advanced topics highlight the broader applications and implications of Pascal's Triangle in various branches of mathematics and science. Specifically, we will discuss Multinomial Coefficients, Pascal's Triangle in Probability, and Pascal's Triangle in Combinatorics, each offering a deeper understanding of this mathematical gem (Anarova et al., 2020).

5.5.1. Multinomial Coefficients

Multinomial coefficients are a natural extension of the binomial coefficients found in Pascal's Triangle, offering a powerful tool for counting and calculating in combinatorics and probability theory. While binomial coefficients are used to count combinations of two items, multinomial coefficients generalize this concept to count combinations of multiple items. In this exposition, we will explore multinomial coefficients, their mathematical expressions, and their significance in various mathematical contexts.

Did You Know?

While binomial coefficients are well-known, there are also multinomial coefficients that come into play when you have more than two categories or choices. They are a powerful tool in statistics, especially in scenarios involving multiple options.

5.5.1.1. Definition and Mathematical Expression

Multinomial coefficients are used to count the number of ways to distribute (n) items into (k) distinct categories or bins, where each category may receive a different number of items. The mathematical expression for a multinomial coefficient, denoted as $C(n; n_1, n_2, \ldots, n_k)$, is given by:

Where:

i. n is the total number of items to be distributed.

ii. k is the number of distinct categories or bins.

iii. $n_1, n_2, \ldots, n_k$ represent the number of items placed in each category, with $(n_1 + n_2 + \ldots + n_k = n)$.

iv. $n!$ is the factorial of (n), which is the product of all positive integers from 1 to (n).

v. $n_i!$ is the factorial of (n_i), representing the product of positive integers from 1 to (n_i).

5.5.1.2. Pascal's Triangle and Multinomial Coefficients

The connection between Pascal's Triangle and multinomial coefficients lies in the way we calculate multinomial coefficients. To compute $C(n; n_1, n_2, \ldots, n_k)$, we can view it as a series of binomial coefficient calculations. Specifically, the formula for multinomial coefficients can be expanded as follows:

$$C(n;n_1,n_2,\ldots,n_k)=\binom{n}{n_1}\cdot\binom{n-n_1}{n_2}\cdot\binom{n-n_1-n_2}{n_3}\cdot\ldots\cdot\binom{n_k}{n_k}$$

This expansion mirrors the process of calculating multinomial coefficients by choosing the number of items to place in each category sequentially. We start by selecting (n_1) items out of (n) to place in the first category, then (n_2) items out of the remaining $(n-n_1)$ for the second category, and so on.

By using binomial coefficients in this manner, we can see that the calculation of multinomial coefficients follows a structured pattern like the expansions in Pascal's Triangle. This connection highlights how Pascal's Triangle serves as a foundation for understanding and calculating more complex combinatorial concepts like multinomial coefficients (Ghent, 1967).

5.5.1.3. Applications in Probability Theory

Multinomial coefficients find extensive applications in probability theory, particularly in problems involving experiments with multiple outcomes or categories. One notable application is the multinomial distribution, which generalizes the binomial distribution to scenarios with more than two possible outcomes. In the multinomial distribution, each outcome is assigned a probability, and the distribution calculates the probabilities of various combinations of outcomes occurring. Multinomial coefficients are used to compute these probabilities efficiently (Uptegrove, 2015).

For example, consider an experiment where a die is rolled (n) times, and we want to find the probability of obtaining (n_1) ones, (n_2) twos, and so on, up to (n_6) sixes. The PMF for this scenario can be expressed using multinomial coefficients:

$$P(X_1=n_1,X_2=n_2,\ldots,X_6=n_6)=\frac{n!}{n_1!\cdot n_2!\cdot\ldots\cdot n_6!}\cdot p_1^{n_1}\cdot p_2^{n_2}\cdot\ldots\cdot p_6^{n_6}$$

Where:

i. $X_1,X_2,\ldots,X_6$ represent the outcomes of rolling the die.

ii. $n_1,n_2,\ldots,n_6$ are the desired counts for each outcome.

iii. n is the total number of rolls.

iv. $p_1, p_2, \ldots, p_6$ are the probabilities of getting each outcome on a single roll.

v. $n_1! \cdot n_2! \cdot \ldots \cdot n_6!$ represents the multinomial coefficient.

Multinomial coefficients facilitate the calculation of probabilities for various outcomes in such experiments, making them an essential tool in probability theory.

5.5.1.4. Applications in Combinatorics

In combinatorics, multinomial coefficients are used to count the number of ways to arrange and distribute objects into multiple categories while accounting for overcounting. This is particularly useful in problems related to permutations and arrangements (Znojil, 2007). For instance, consider a problem where we want to arrange the letters in the word "MISSISSIPPI."

Here, we have $(n=11)$ letters, and there are four categories: M, I, S, and P. The multinomial coefficient $C(11;1,4,4,2)$ calculates the number of distinct arrangements, accounting for the repetition of letters.

Generalizations of multinomial coefficients also play a role in combinatorial identities and combinatorial analysis. They appear in formulas related to partitions of integers, counting restricted permutations, and solving combinatorial problems involving various restrictions.

5.5.1.5. Mathematical Significance and Complexity

Multinomial coefficients offer a deeper level of combinatorial understanding and can be seen as building upon the foundation laid by binomial coefficients in Pascal's Triangle. They provide a systematic way to count arrangements and distributions of items into multiple categories, which is essential in probability theory and combinatorics.

However, it is important to note that as the number of categories (k) or the values of (n_i) become larger, calculating multinomial coefficients directly can be computationally complex due to the factorial terms involved. In such cases, techniques like recursive algorithms or efficient mathematical software are employed to compute these coefficients (Brietzke, 2008).

5.5.2. Pascal's Triangle in Probability

Probability theory is a branch of mathematics that deals with uncertainty and randomness, and it has a deep connection with Pascal's Triangle. Pascal's Triangle provides a versatile framework for understanding and solving probability problems, especially those involving combinations, permutations, and distributions. In this exposition, we will explore the role of Pascal's Triangle in probability theory, highlighting its applications in various probabilistic scenarios.

5.5.2.1. Binomial Distribution and Probability Mass Functions

The most direct connection between Pascal's Triangle and probability lies in the binomial distribution. The binomial distribution describes the probability of obtaining a specific number of successes in a fixed number of independent Bernoulli trials, where each trial has only two possible outcomes: success or failure.

Pascal's Triangle plays a pivotal role in calculating the probabilities associated with the binomial distribution. Specifically, the binomial coefficient $C(n,k)$ from Pascal's Triangle is used to compute the PMF of the binomial distribution (Batanero & Parzysz, 2005).

The PMF of the binomial distribution is given by:

$$P(X=k)=C(n,k)\cdot p^k \cdot q^{n-k}$$

Where:

i. X represents the random variable denoting the number of successes.

ii. k is the specific number of successes we are interested in.

iii. n is the total number of trials.

iv. p is the probability of success on each trial.

v. q is the probability of failure on each trial $(q=1-p)$.

The binomial coefficient $C(n,k)$ in this formula corresponds to the kth term in the row of Pascal's Triangle. This coefficient captures the number of ways to choose (k) successes out of (n) trials, making it an essential component of the binomial distribution.

In probability theory, this expansion is used to calculate probabilities in scenarios where there are two possible outcomes (success and failure) for each trial, and we want to know the probability of getting a specific number of successes in (n) trials. By substituting $(a=p)$ (probability of success) and $(b=q)$ (probability of failure), the formula becomes:

$$(p+q)^n = \sum_{k=0}^{n} C(n,k) \cdot p^{n-k} \cdot q^k$$

This expansion gives the probability distribution of the binomial random variable (X), which represents the number of successes in (n) independent Bernoulli trials (Pain & Poirier, 2020).

5.5.2.2. Applications in Bernoulli Trials

Pascal's Triangle is instrumental in solving problems related to Bernoulli trials, which are experiments with two possible outcomes: success and failure. These trials are characterized by the probability of success (p) and the probability of failure $(q=1-p)$. Some common applications include:

1. **Coin Flips:** When flipping a fair coin (where $(p=0.5)$ multiple times, Pascal's Triangle can be used to find the probability of obtaining a certain number of heads or tails.
2. **Biased Dice Rolls:** In situations where dice or other random devices are biased, Pascal's Triangle helps determine the likelihood of obtaining specific outcomes.
3. **Success-Failure Experiments**: In scenarios where an event can result in success or failure (e.g., manufacturing quality control, medical diagnosis), Pascal's Triangle assists in calculating probabilities associated with different numbers of successes.
4. **Probability of Winning a Game:** When playing games with multiple rounds or trials (e.g., card games), Pascal's Triangle aids in calculating the probability of winning a certain number of rounds (Gould, 1969).

5.5.2.3. Hypergeometric Distribution

Beyond the binomial distribution, Pascal's Triangle also plays a role in understanding the hypergeometric distribution. The hypergeometric distribution models situations where there is a finite population, and we are interested in the probability of obtaining a specific number of successes without replacement. It arises in problems such as sampling without replacement from a finite deck of cards, drawing balls from urns, or selecting items from a limited population. The PMF of the hypergeometric distribution involves binomial coefficients and can be expressed as:

$$P(X=k)=\frac{C(N_1,k)\cdot C(N_2,n-k)}{C(N,n)}$$

Where:

i. X is the random variable representing the number of successes

ii. k is the specific number of successes.

iii. N_1 is the number of successful items in the population.

iv. N_2 is the number of unsuccessful items in the population.

v. N is the total population size.

vi. n is the number of items drawn from the population.

The binomial coefficients $C(N_1,k), C(N_2,n-k),$ and $C(N,n)$ are related to Pascal's Triangle and reflect the combinatorial aspects of the hypergeometric distribution (Provost & Ratemi, 2011).

5.5.2.4. Combinatorial Interpretations

Pascal's Triangle offers combinatorial interpretations of probabilities in various scenarios. The hypergeometric distribution relies on the concept of combinations without replacement, whereas Pascal's Triangle provides insights into the counting of successful and unsuccessful items events (Allouche & Johnson, 2018).

5.5.2.5. Mathematical Significance

The significance of Pascal's Triangle in probability lies in its ability to simplify complex probabilistic calculations by providing a systematic way to compute binomial coefficients and probabilities associated with binomial and hypergeometric distributions. This mathematical tool is fundamental in both theoretical and applied probability, making Pascal's Triangle an indispensable asset for probabilistic modeling and analysis.

5.5.3. Pascal's Triangle in Combinatorics

Combinatorics, a branch of mathematics concerned with counting, arrangements, and combinations, finds a rich source of inspiration and utility in Pascal's Triangle. This triangular array of numbers provides profound insights into combinatorial concepts, facilitating the enumeration of various combinatorial objects and the solution of complex counting problems (Vincenzi & Siani, 2014).

5.5.3.1. Combinatorial Interpretations

Pascal's Triangle offers combinatorial interpretations for its elements, making it a versatile tool in combinatorics. Each number in the triangle can be associated with a specific combinatorial problem or counting task. These interpretations provide a direct link between the numbers in the triangle and various combinatorial objects. Some notable combinatorial interpretations include (Anarova et al., 2020) (Figure 5.12):

1. **Combinations with Repetition:** Pascal's Triangle can be extended to count combinations with repetition. For example, the third diagonal of the triangle represents the triangular numbers, which count the number of ways to distribute identical items into distinct boxes.
2. **Catalan Numbers:** The diagonal sums of Pascal's Triangle lead to the Catalan numbers, a famous sequence in combinatorics. These numbers count various combinatorial objects, such as balanced parentheses expressions, non-crossing partitions, and binary trees.
3. **Compositions and Partitions:** Different paths through Pascal's Triangle correspond to compositions and partitions of numbers. For example, the paths along the diagonals represent compositions of natural numbers, while paths along the rows represent partitions of natural numbers (AP Stakhov, 2005).

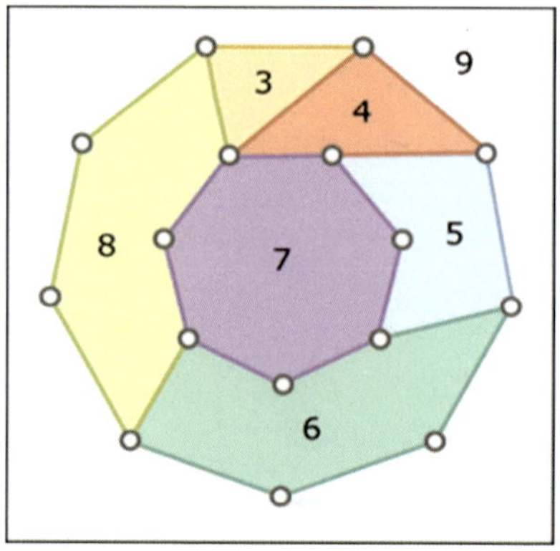

Figure 5.12. Schematic of combinatorial analysis.

Source: Prof. Richard Stanley, Creative Commons License.

5.5.3.2. Counting and Enumerating Combinatorial Objects

Pascal's Triangle is a valuable tool for counting and enumerating various combinatorial objects. It simplifies the process of determining the number of ways to arrange, select, or partition elements in different configurations. Some examples of combinatorial objects that can be counted using Pascal's Triangle include:

1. **Subsets and Submultisets:** Pascal's Triangle can be used to count the number of subsets (combinations) of a set, as well as the number of ways to distribute identical items into distinct categories (submultisets).
2. **Permutations:** When arranging objects in a specific order, Pascal's Triangle can count permutations with and without repetitions, allowing us to find the number of distinct arrangements of elements.
3. **Paths and Trees:** The Catalan numbers derived from Pascal's Triangle are instrumental in counting various types of paths (e.g., Dyck paths) and trees (e.g., binary trees, rooted trees).
4. **Partitions:** Pascal's Triangle assists in counting partitions of numbers into distinct parts or into parts with restrictions (Bell & Terry, 2018).
5. **Compositions:** It aids in counting compositions, which are representations of numbers as sums of positive integers.

5.5.3.3. Pascal's Identity

Pascal's Triangle also gives rise to Pascal's Identity, a fundamental result in combinatorics that establishes a relationship between binomial coefficients. This identity is expressed as:

$$C(n,k) = C(n-1,k-1) + C(n-1,k)$$

Pascal's Identity demonstrates how the combination of elements in Pascal's Triangle results in the formation of new combinations. It is particularly useful for recursive counting problems, where the number of combinations at one level depends on the combinations at the previous level. This recursive property of Pascal's Triangle simplifies the enumeration of various combinatorial objects and is a powerful tool in combinatorics (Ventresca, 2012).

5.5.3.4. Mathematical Significance

The mathematical significance of Pascal's Triangle in combinatorics cannot be overstated. It serves as a unifying framework for understanding and solving a wide range of counting problems, from basic combinations to intricate combinatorial structures. Its combinatorial interpretations provide intuitive explanations for the numbers within the triangle, making complex counting tasks more accessible and approachable. Pascal's Triangle bridges the gap between algebraic expressions and combinatorial outcomes, demonstrating the inherent connections between mathematical structures (Saliba, 2019).

5.5.4. Combinatorial Problems

The following are combinatorial problems:

5.5.4.1. Counting Combinations

Combinatorics is the study of counting and arranging objects, and Pascal's Triangle simplifies the counting of combinations. Combinations are subsets of a set, and calculating the number of ways to select (k) items from a set of (n) items without regard to order is a common combinatorial problem.

Pascal's Triangle provides a direct method for calculating binomial coefficients $C(n,k)$, which represents the number of combinations. The value at row (n) and position (k) in the triangle is precisely $C(n,k)$, making it easy to determine the number of combinations for various values of (n) and (k).

5.5.4.2. Permutations with and without Repetition

Permutations involve arranging objects in a specific order. Pascal's Triangle assists in counting permutations both with and without repetition (Powell, 2003).

For permutations without repetition, where all objects are distinct, the number of permutations is $(n!)$, the factorial of (n). For permutations with repetition, where some objects may be identical, the binomial coefficient $C(n,k_1,k_2,\ldots,k_r)$ represents the number of permutations of (n) objects with \(r\) distinct types, where (k_i) objects are of type (i). This concept applies to various scenarios, such as arranging letters in a word or arranging items in a sequence (Leroy et al., 2016).

5.5.4.3. Counting Paths and Trees

Pascal's Triangle finds applications in counting paths and trees in combinatorial problems. Paths and trees are fundamental structures in combinatorics, often used to model various real-world situations.

1. **Dyck Paths:** Dyck paths, which represent valid combinations of open and closed parentheses, are counted using the Catalan numbers derived from diagonal sums in Pascal's Triangle. These paths have applications in various combinatorial problems, including parsing in computer science and RNA secondary structure prediction in biology.
2. **Binary Trees:** The Catalan numbers also count binary trees, which are used to represent hierarchical structures in computer science and data structures. Pascal's Triangle aids in determining the number of different binary trees with a given number of nodes.

5.5.4.4. Partitions and Compositions

Partitions and compositions involve representing numbers as sums or combinations of smaller numbers (Comtet, 1974). These concepts are essential in combinatorial mathematics, and Pascal's Triangle provides insights into their enumeration.

1. **Partitions:** Partitions of a number into distinct parts are counted using the numbers along the rows of Pascal's Triangle. For example, the number of ways to partition a number into distinct positive integers can be determined using the values in Pascal's Triangle.
2. **Compositions:** Compositions, which represent numbers as sums of positive integers, are related to paths in Pascal's Triangle. The different paths through the triangle correspond to various compositions of a number.

5.5.5. Algebraic Manipulation

The following are the algebraic manipulations:

5.5.5.1. Algebraic Identities and Combinatorial Connections

Pascal's Triangle also reveals algebraic identities that have combinatorial interpretations. One such identity is Pascal's Identity:

$$C(n,k)=C(n-1,k-1)+C(n-1,k)$$

This identity relates binomial coefficients and expresses how combinations at one level depend on combinations at the previous level. While it is a combinatorial identity, it has applications in algebraic manipulations, particularly in simplifying and expanding expressions involving binomial coefficients. Pascal's Triangle's capacity to establish connections between algebraic identities and combinatorial objects highlights its versatility as a mathematical tool. It allows mathematicians and scientists to seamlessly transition between algebraic and combinatorial aspects of mathematics.

5.5.5.2. Applications in Algebraic Expressions

Beyond probability and combinatorial problems, Pascal's Triangle finds applications in algebraic expressions. For instance, when raising a binomial expression $(a+b)^n$ to a high power, the coefficients in the expansion can be efficiently determined using Pascal's Triangle. This simplifies complex algebraic manipulations and polynomial expansions, saving time and effort in mathematical computations (Stigler, 2014).

Activity 5.1

Imagine you're planning seating arrangements for a wedding reception with 10 guests. You have two long tables, and you want to know how many different ways you can seat your guests. How does the concept of binomial coefficients relate to solving this seating arrangement challenge?

5.6. SUMMARY

This chapter begins with an introduction to Pascal's Triangle, highlighting its remarkable mathematical properties and the method for constructing it. The chapter explains the patterns within Pascal's Triangle, demonstrating its significance as a mathematical tool. The focus then shifts to binomial coefficients, offering a clear definition and emphasizing their role in combinatorics. The chapter delves into various aspects of binomial coefficients, including notation, symbolism, and combinatorial interpretations. It explains how binomial coefficients are vital in binomial expansion and their applications in algebra. The properties of Pascal's Triangle are thoroughly examined, highlighting its symmetry, sums, differences, and diagonal sums. Advanced topics, such as multinomial coefficients, probability applications, combinatorics, and algebraic manipulations, broaden the understanding of this mathematical concept.

REFERENCES

1. Allen, M. A., (2022). *On a Two-Parameter Family of Generalizations of Pascal's Triangle*. arXiv preprint arXiv:2209.01377.
2. Allenby, R. B., & Slomson, A., (2010). *How to Count: An Introduction to Combinatorics* (Vol. 1, No. 2, pp. 2–10). CRC Press.
3. Allouche, J. P., & Johnson, T., (2018). Combinatorics of words and morphisms in some pieces of Tom Johnson. *Journal of Mathematics and Music, 12*(3), 248–257.
4. Al-Najjar, S. A. S., (2015). Computing the number of integral points in4-dimensional ball using Tutte polynomial. *Engineering and Technology Journal, 33*(8 Part (B) Scientific).
5. Alper, P., & Yang, Y., (1999). Tea tasting and Pascal triangles. *Humanistic Mathematics Network Journal, 1*(21), 20.
6. Anarova, S. A., Ibrohimova, Z., Narzulloyev, O., & Qayumova, G., (2020). *Mathematical Modeling of Pascal Triangular Fractal Patterns and its Practical Application* (Vol. 2, No. 1, pp. 5–25). Paper presented at the International Conference on Intelligent Human Computer Interaction.
7. Annamalai, C., (2022a). *Computation and Combinatorial Techniques for Binomial Coefficients and Geometric Series*, *1*(3), 1–12.
8. Annamalai, C., (2022b). *Extension of Binomial Series with Optimized Binomial Coefficient*. Available at SSRN 4129875.
9. Bardzell, M., & Poimenidou, E., (2011). Binomial coefficients modulo a prime—A visualization approach to undergraduate research. *PRIMUS, 21*(3), 238–251.
10. Batanero, C., Henry, M., & Parzysz, B., (2005). The nature of chance and probability. In: *Exploring Probability in School: Challenges for Teaching and Learning* (Vol. 1, pp. 15–37). Springer.
11. Beiu, V., Dăuş, L., Jianu, M., Mihai, A., & Mihai, I., (2022). On a surface associated with Pascal's triangle. *Symmetry, 14*(2), 411.
12. Bell, T., Frankson, S., Sachdeva, N., & Terry, M., (2018). *The Relationship Between Pascal's Triangle and Random Walks.* arXiv preprint arXiv:1811.02708.
13. Bober, Z., & Ignatyev, M. B., (2017). System analysis of problems of the family on the basis of the linguo-combinatorial simulation. *System Analysis in Projection and Management*, 1, 143.

14. Brietzke, E. H., (2008). An identity of Andrews and a new method for the Riordan array proof of combinatorial identities. *Discrete Mathematics, 308*(18), 4246–4262.
15. Cameron, P. J., (1994). *Combinatorics: Topics, Techniques, Algorithms* (Vol. 1, No. 4, pp. 33–49). Cambridge University Press.
16. Clark, D. S., (1982). On some abstract properties of binomial coefficients. *The American Mathematical Monthly, 89*(7), 433–443.
17. Cobeli, C., & Zaharescu, A., (2013). Promenade around Pascal triangle—Number motives. *Bulletin Mathématique de la Société des Sciences Mathématiques de Roumanie, 2*(2), 73–98.
18. Comtet, L., (1974). *Advanced Combinatorics: The Art of Finite and Infinite Expansions* (Vol. 2, No. 5, pp. 5–13). Springer Science & Business Media.
19. De Weger, B. M., (1997). Equal binomial coefficients: Some elementary considerations. *Journal of Number Theory, 63*(2), 373–386.
20. Doh, P. K., Adjallah, K. H., & Birregah, B., (2021). Thirty-six full matrix forms of the Pascal triangle: Derivation and symmetry relations. *Scientific African, 13*, e00932.
21. Donnelly, R. G., Dunkum, M. W., George, C., & Schnake, S., (2018). *Counting Odd Numbers in Truncations of Pascal's Triangle.* arXiv preprint arXiv:1807.08181.
22. Edwards, A. W. F., (2019). *Pascal's Arithmetical Triangle: The Story of a Mathematical Idea* (Vol. 2, No. 5, pp. 7–13). Courier Dover Publications.
23. Edwards, A., (2001). Blaise Pascal: B. 19 June 1623 d. 19 August 1662. In: *Statisticians of the Centuries* (Vol. 1, pp. 17–22). Springer.
24. English, L. D., (2005). Combinatorics and the development of children's combinatorial reasoning. In: *Exploring Probability in School: Challenges for Teaching and Learning* (Vol. 2, pp. 121–141). Springer.
25. Feng, J., Shi, C., & Zhao, H., (2019). *Two Deformed Pascal's Triangles and Its New Properties*. arXiv preprint arXiv:1909.00611.
26. Flusser, P., & Francia, G. A., (2000). Derivation and visualization of the binomial theorem. *International Journal of Computers for Mathematical Learning, 5*, 3–24.
27. Fowler, D., (1996). The binomial coefficient function. *The American Mathematical Monthly, 103*(1), 1–17.

28. Fraenkel, A. S., & Kontorovich, A., (2007). The Sierpiński sieve of Nim-varieties and binomial coefficients. In: *Combinatorial Number Theory* (Vol. 1, pp. 209–227). De Gruyter Berlin.

29. Gandhi, M. A., & Padi, T. R., (2023) *Combinatorial Probability Distribution for 2nd Diagonal Elements of the Pascal's Triangle with Sequence of Natural Numbers in Prime Diagonal, 3*(5), 10–18.

30. García-García, J. I., Fernández, C. N. A., Arredondo, E. H., & Imilpán, R. I. A., (2022). The binomial distribution: Historical origin and evolution of its problem situations. *Mathematics, 10*(15), 2680.

31. Ghent, A. W., (1967). Selected problems in biometry part 2. The structure of the binomial distribution. *Bios*, 115–134.

32. Gould, H., (1969). The bracket function and Fontené-Ward generalized binomial coefficients with application to fibonomial coefficients. *Fibonacci Quart, 7*, 23–40.

33. Granville, A., (1997). Arithmetic properties of binomial coefficients. I. Binomial coefficients modulo prime powers. *Organic Mathematics (Burnaby, BC, 1995), 20*, 253–276.

34. Gwetta, E., Pacurar, A., & Smith, E. M., (2012). A generalization of Pascal's triangle. *DePaul Discoveries, 1*(1), 13.

35. Hari, B. K., (2011). *Probability Theory and Stochastic Processes for Engineers* (Vol. 2, No. 4, pp. 1–25). Pearson Education India.

36. Hilton, P., & Pedersen, J., (1987). Looking into Pascal's triangle: Combinatorics, arithmetic, and geometry. *Mathematics Magazine, 60*(5), 305–316.

37. Holte, J. M., (1997). Carries, combinatorics, and an amazing matrix. *The American Mathematical Monthly, 104*(2), 138–149.

38. Horn, M. E., (2006). *The Didactical Relevance of the Pauli Pascal Triangle (Die Didaktische Relevanz des Pauli-Pascal-Dreiecks).* arXiv preprint physics/0611277.

39. Huard, J. G., Spearman, B. K., & Williams, K. S., (1998). Pascal's triangle (mod 8). *European Journal of Combinatorics, 19*(1), 45–62.

40. Islam, M. S., Islam, M. R., Hossan, M. S., & Kibria, M. H., (2022). Generating binomial coefficients in a row of Pascal's triangle from extensions of powers of eleven. *Heliyon, 8*(11).

41. Konvalina, J., (2000). A unified interpretation of the binomial coefficients, the Stirling numbers, and the Gaussian coefficients. *The*

American Mathematical Monthly, 107(10), 901–910.

42. Kuzmin, O., & Starkov, B., (2021). *Application of Hierarchical Structures Based on Binary Matrices with the Generalized Arithmetic of Pascal's Triangle in Route Building Problems* (Vol. 1, No. 3, pp. 1–15). Paper presented at the Journal of Physics: Conference Series.
43. Laudański, L. M., & Laudański, L. M., (2013). Binomial distribution. *Between Certainty and Uncertainty: Statistics and Probability in Five Units with Notes on Historical Origins and Illustrative Numerical Examples,* 87–127.
44. Lay-Yong, L., (1980). The Chinese connection between the Pascal triangle and the solution of numerical equations of any degree. *Historia Mathematica, 7*(4), 407–424.
45. Leavitt, S., (2011). *The Simple Complexity of Pascal's Triangle* (Vol. 2, No. 23, pp. 10–15). Master Project, Department of Mathematics, Penn State College of Information Science and Technology, Pennsylvania, Westgate Building University Park, United State.
46. Leroy, J., Rigo, M., & Stipulanti, M., (2016). Generalized Pascal triangle for binomial coefficients of words. *Advances in Applied Mathematics, 80*, 24–47.
47. Liu, C. S., & Kuo, C. L., (2016). A multiple-scale Pascal polynomial triangle solving elliptic equations and inverse Cauchy problems. *Engineering Analysis with Boundary Elements, 62*, 35–43.
48. Lovasz, L., Pelikan, J., Vesztergombi, K., Lovász, L., Pelikán, J., & Vesztergombi, K., (2003). Binomial coefficients and Pascal's triangle. *Discrete Mathematics: Elementary and Beyond*, 43–64.
49. Lovász, L., Pelikán, J., Vesztergombi, K., Lovász, L., Pelikán, J., & Vesztergombi, K., (2003). Combinatorial probability. *Discrete Mathematics: Elementary and Beyond*, 77–85.
50. Lu, T., Zhang, P., & Li, H., (2019). *Practice Teaching Reform of Discrete Mathematics Model Based on D3.js* (Vol. 1, No. 3, pp. 12–30). Paper presented at the 2019 14th International Conference on Computer Science & Education (ICCSE).
51. Murodillayevich, N. F., Amanbayevna, A. S., & Mirzayevich, N. O., (2019). *Mathematical and Software of Fractal Structures from Combinatorial Numbers* (Vol. 89, No. 7, p. 433). Paper presented at the 2019 International Conference on Information Science and Communications Technologies (ICISCT).

52. Mustafa, A. N., (2020). History of mathematics education in China. *Pasundan Journal of Mathematics Education Jurnal Pendidikan Matematika, 10*(2), 1–22.
53. Nelson, R., (2013). *Probability, Stochastic Processes, and Queueing Theory: The Mathematics of Computer Performance Modeling* (Vol. 1, No. 2, pp. 1–27). Springer Science & Business Media.
54. Neuwirth, E., (1998). *Spreadsheets: Just Smart Calculators or a New Paradigm for Thinking About Mathematical Structures?* (Vol. 1, No. 5, pp. 1–22). Paper presented at the Information and Communications Technologies in School Mathematics: IFIP TC3/WG3. 1 Working Conference on Secondary School Mathematics in the World of Communication Technology: Learning, Teaching, and the Curriculum, 26–31 October 1997, Grenoble, France.
55. Nolan, C., (2003). Combinatorial space in nineteenth-and early twentieth-century music theory. *Music Theory Spectrum, 25*(2), 205–241.
56. Pain, J. C., & Poirier, M., (2020). Analytical and numerical expressions for the number of atomic configurations contained in a supershell. *Journal of Physics B: Atomic, Molecular and Optical Physics, 53*(11), 115002.
57. Pengelley, D., (2009) Pascal's Treatise on the arithmetical triangle: Mathematical induction, combinations. *Resources for Teaching Discrete Mathematics, 1*(185), 11–19.
58. Powell, A. B., (2003). *"So Let's Prove It!": Emergent and Elaborated Mathematical Ideas and Reasoning in the Discourse and Inscriptions of Learners Engaged in a Combinatorial Task* (Vol. 1, No. 2, pp. 1–15). Rutgers The State University of New Jersey, School of Graduate Studies.
59. Provost, S. B., & Ratemi, W. M., (2011). Polynomial expansions via embedded Pascal's triangles. *Acta et Commentationes Universitatis Tartuensis de Mathematica, 15*(1), 45–60.
60. Ratemi, W. M., (2012). *Embedded Pascal Triangles and Its Application for Minimal Cut Sets of Fault Tree Analysis* (Vol. 1, No. 2, pp. 1–20). Paper presented at the Abstract-2nd International Conference on Mathematical Applications in Engineering.
61. Rogers, D., (2006). *A Schröder Triangle: Three Combinatorial Problems.* Paper presented at the Combinatorial Mathematics V:

Proceedings of the Fifth Australian Conference, Held at the Royal Melbourne Institute of Technology.

62. Rushdi, A. M. A., & Al-Amoudi, M. A. R., (2019). Recursive computation of binomial and multinomial coefficients and probabilities. *Advances in Mathematics and Computer Science 1*, 113–129.
63. Russell, G., (2022). From towers of linking cubes to the binomial expansion theorem: What can be learned about combinatorics? *Canadian Journal of Science, Mathematics and Technology Education, 22*(3), 700–713.
64. Saliba, M., (2019). Polyelemental, multicomponent perovskite semiconductor libraries through combinatorial screening. *Advanced Energy Materials, 9*(25), 1803754.
65. Scheaffer, R. L., Watkins, A. E., & Landwehr, J. M., (1998). What every high-school graduate should know about statistics. *Reflections on Statistics: Learning, Teaching, and Assessment in Grades K-12*, 3–31.
66. Sills, A. V., (2023). A refinement of the binomial distribution using the quantum binomial theorem. *Communications in Statistics-Theory and Methods, 52*(2), 294–308.
67. Skurnick, R., Baderian, A., & Martin, J., (2012). A note on the extended binomial coefficient for negative integers and an extension of Pascal's triangle. *Mathematics and Computer Education, 46*(1), 35.
68. Smith, K. J., (1973). Pascal's triangle. *The Two-Year College Mathematics Journal, 4*(1), 1–13.
69. Sofo, A., (2009). Sums of derivatives of binomial coefficients. *Advances in Applied Mathematics, 42*(1), 123–134.
70. Song, Y., & Deng, Y., (2021). Entropic explanation of power set. *International Journal of Computers, Communications & Control, 16*(4), 4413.
71. Spresser, D. M., & Fritz, J. S., (1996). Two combinatorial identities derived from the recurrence relation. *International Journal of Mathematical Education in Science and Technology, 27*(1), 23–27.
72. Stakhov, A., (2005). The generalized principle of the golden section and its applications in mathematics, science, and engineering. *Chaos, Solitons & Fractals, 26*(2), 263–289.

73. Stakhov, A., (2007). The generalized golden proportions, a new theory of real numbers, and ternary mirror-symmetrical arithmetic. *Chaos, Solitons & Fractals, 33*(2), 315–334.
74. Stigler, S., (2014). Soft questions, hard answers: Jacob Bernoulli's probability in historical context. *International Statistical Review, 82*(1), 1–16.
75. Uptegrove, E. B., (2015). Shared communication in building mathematical ideas: A longitudinal study. *The Journal of Mathematical Behavior, 40*, 106–130.
76. Van, D. F. C., Ipolitti, G., Nicolau, Jr. D. V., Sudalaiyadum, P. A., Kašpar, O., Kheireddine, S., & Nicolau, D. V., (2018). Something has to give: Scaling combinatorial computing by biological agents exploring physical networks encoding NP-complete problems. *Interface Focus, 8*(6), 20180034.
77. Ventresca, M., (2012). Global search algorithms using a combinatorial unranking-based problem representation for the critical node detection problem. *Computers & Operations Research, 39*(11), 2763–2775.
78. Vincenzi, G., & Siani, S., (2014). Fibonacci-like sequences and generalized Pascal's triangles. *International Journal of Mathematical Education in Science and Technology, 45*(4), 609–614.
79. Wright, T., (1995). Pascal's triangle gets its genes from Stirling numbers of the first kind. *The College Mathematics Journal, 26*(5), 368–371.
80. Yurkin, A. V., (2013). *Ray Trajectories, Binomial of a New Type, and the Binary System; on Binomial Distribution of the Second (Nonlinear) Type for Big Binomial Power.* arXiv preprint arXiv:1302.4842.
81. Zhang, Y., Xiao, H., Chen, D., Ding, S., & Chen, P., (2016). *Link from ZD Control to Pascal's Triangle Illustrated via Multiple-Integrator Systems* (Vol. 1, No. 2, pp. 1–25). Paper presented at the 2016 12th International Conference on Natural Computation, Fuzzy Systems and Knowledge Discovery.
82. Znojil, M., (2007). Tridiagonal-symmetric N-by-N Hamiltonians and a fine-tuning of their observability domains in the strongly non-Hermitian regime. *Journal of Physics A: Mathematical and Theoretical, 40*(43), 13131.

CHAPTER 6

Recurrence Relations and Recursive Sequences

Contents

6.1. INTRODUCTION

Recurrence relations are a fundamental concept in mathematics that describes sequences or progressions where each term depends on one or more previous terms. They provide a powerful framework for modeling and analyzing various phenomena in mathematics and the real world. In this chapter, we will delve into the definition and notation of recurrence relations, explore examples of their application in real-world problems, and distinguish between linear and nonlinear recurrence relations (Berthomieu & Faugère, 2016). A recurrence relation is a mathematical equation or formula that defines the relationship between the terms of a sequence by expressing each term in terms of one or more preceding terms. It is typically represented using mathematical notation. Consider a sequence (a_n) where (n) is a positive integer. A first-order recurrence relation for this sequence can be defined as:

$$a_{n+1} = f(a_n)$$

In this equation, (a_{n+1}) represents the next term in the sequence, and $f(a_n)$ is a function that depends on the current term (a_n). This basic form illustrates the recursive nature of recurrence relations, where each term relies on the previous one (Stoll, 1997).

Let's examine a simple example to illustrate this concept. Suppose we have a sequence of positive integers defined as follows:

$$a_1 = 1$$

$$a_{n+1} = a_n + 2$$

In this recurrence relation, each term (a_{n+1}) is determined by adding 2 to the previous term (a_n). This simple relation generates the sequence $(1, 3, 5, 7, \ldots)$, where each term is 2 greater than the previous term (Shannon, 1974).

Recurrence relations find extensive applications in various real-world problems, from modeling population growth to analyzing financial investments. These mathematical tools allow us to describe how quantities change over time or in response to specific conditions. Here are a few examples of real-world problems that can be modeled using recurrence relations:

1. **Population Growth:** The population growth model is a classic example of a recurrence relation used in biology and ecology to describe how a population's size changes over time. It considers factors such as birth rate, death rate, and carrying capacity to model population dynamics (Martins et al., 2016).
2. **Fibonacci Sequence:** The Fibonacci sequence is a well-known recurrence relation that appears in diverse fields, including biology, art, and computer science. It describes the growth of a population of rabbits over discrete time intervals, where each term is the sum of the two preceding terms: $(F_{n+2} = F_{n+1} + F_n)$.
3. **Financial Investments:** Recurrence relations are applied in finance to model the growth of investments or savings over time. The future value of an investment can be calculated recursively based on the initial investment, interest rate, and compounding frequency.
4. **Dynamic Programming:** In computer science and optimization problems, dynamic programming algorithms often rely on recurrence relations to find optimal solutions by breaking down complex problems into simpler subproblems (Kessler & Anderson, 1986).

Recurrence relations can be categorized into linear and nonlinear types based on their mathematical structure and behavior. Understanding this distinction is essential for solving and analyzing recurrence relations effectively.

1. **Linear Recurrence Relations:** In a linear recurrence relation, each term of the sequence is expressed as a linear combination of previous terms, typically involving constant coefficients. The general form of a linear recurrence relation of order (k) is:

$$a_{n+k} = c_1 a_{n+k-1} + c_2 a_{n+k-2} + \ldots + c_k a_n$$

Where $(c_1, c_2, \ldots, c_k)$ are constants. Linear recurrence relations exhibit predictable and well-behaved solutions. The Fibonacci sequence, for example, is a linear recurrence relation.

2. **Nonlinear Recurrence Relations:** Nonlinear recurrence relations involve terms that are not expressed as linear combinations of previous terms. They can have more complex functional relationships between terms. One common example is the logistic map, which models chaotic behavior in dynamic systems (Torres et al., 2005).

To illustrate the difference between linear and nonlinear recurrence relations, let's consider two recurrence relations:

- **Linear Example:**

The recurrence relation for a simple exponential growth sequence is linear:

$$a_{n+1} = r \cdot a_n$$

Here, each term (a_{n+1}) is obtained by multiplying the previous term (a_n) by a constant (r).

- **Nonlinear Example:**

The logistic map, a classic nonlinear recurrence relation, is defined as:

$$a_{n+1} = r \cdot a_n (1 - a_n)$$

In this case, the next term (a_{n+1}) depends on the previous term (a_n) in a nonlinear manner involving multiplication and subtraction (Ash, 1986).

Understanding whether a recurrence relation is linear or nonlinear is crucial for selecting appropriate solution methods and analyzing the behavior of the sequence it generates. Hence, recurrence relations provide a mathematical framework for modeling sequences and progressions in various contexts. They are expressed using mathematical notation, find practical applications in real-world problems, and can be categorized as linear or nonlinear based on their mathematical structure. In the following sections, we will delve deeper into solving and analyzing recurrence relations of different types and complexities (Zhou et al., 2002).

6.2. FIRST-ORDER RECURRENCE RELATIONS

In this section, we will study the first-order recurrence relations. These mathematical constructs play a crucial role in modeling and understanding sequences where each term depends directly on the previous one. We explore the derivation and solutions of first-order recurrence relations, highlighting their fundamental properties and applications. Additionally, we examine how first-order recurrence relations relate to concepts like exponential growth and decay, as well as explore scenarios involving non-homogeneous relations, which introduce external influences into the sequence's evolution.

6.2.1. Derivation and Solutions in Combinatorics

In combinatorics, first-order recurrence relations serve as valuable tools for counting and enumerating various combinatorial structures. These relations help us understand how the number of objects or configurations changes from one step to the next. In this section, we will explore the derivation and solutions of first-order recurrence relations in the context of combinatorial counting problems (Pan et al., 2023).

Remember

First-order recurrence relations are often encountered in combinatorics and can describe exponential growth or decay processes.

6.2.1.1. Derivation of First-Order Recurrence Relations

A first-order recurrence relation in combinatorics typically arises when we are counting objects or arrangements, and each step or term depends on the previous one. To derive such a relation, we often employ combinatorial principles and logical reasoning.

- **Example:**

Counting the number of ways to arrange (n) distinct objects in a row. Let (a_n) represent the number of such arrangements for (n) objects. To derive a first-order recurrence relation for (a_n), we can break down the problem into smaller cases based on the position of the first object in the arrangement.

There are (n) possibilities for the first object's position.

1. If the first object is placed in the first position, we have (a_{n-1}) ways to arrange the remaining $(n-1)$ objects.
2. If the first object is placed in the second position, we still have (a_{n-1}) ways to arrange the remaining $(n-1)$ objects.

This pattern continues for each possible position of the first object. Therefore, the total number of arrangements for (n) objects can be expressed as:

$$a_n = n \cdot a_{n-1}$$

This is a first-order recurrence relation that relates (a_n) to (a_{n-1}) in a combinatorial context.

6.2.1.2. Solutions of First-Order Recurrence Relations

Solving first-order recurrence relations in combinatorics involves finding a closed-form expression for the sequence (a_n) in terms of (n). This closed-form solution simplifies calculations and allows us to determine the number of arrangements for any given value of (n).

The first-order recurrence relation $(a_n = n \cdot a_{n-1})$ can be solved iteratively or algebraically. Iteratively, we can start with an initial condition (usually (a_0) or (a_1) and calculate subsequent terms using the relation (Kara & Ilkhan, 2016). Algebraically, we can write the relation in a more standard form:

$$\frac{a_n}{a_{n-1}} = n$$

To find a closed-form solution, we can take the product of both sides of this equation for all values of (n) from 2 to (n):

$$\frac{a_n}{a_{n-1}} \cdot \frac{a_{n-1}}{a_{n-2}} \cdot \ldots \cdot \frac{a_2}{a_1} = 2 \cdot 3 \cdot \ldots \cdot n$$

Notice that the left side telescopes, simplifying to:

$$\frac{a_n}{a_1} = n!$$

Solving for (a_n), we obtain the closed-form solution:

$a_n = n!$

This solution represents the number of ways to arrange (n) distinct objects in a row, confirming the combinatorial interpretation of the recurrence relation. In combinatorics, deriving and solving first-order recurrence relations allows us to efficiently count and enumerate various combinatorial structures. These relations provide a systematic approach to understanding and solving counting problems, making them a valuable tool in the field of mathematics (Viswanath, 2000).

6.2.2. Exponential Growth and Decay

In first-order recurrence relations, one of the most prevalent and influential concepts is exponential growth and decay. This mathematical phenomenon manifests when a quantity undergoes continuous change over discrete time intervals, and each step involves a fixed proportional increase (growth) or decrease (decay) relative to the previous value.

6.2.2.1. Exponential Growth

Exponential growth is a mathematical model that describes the rapid increase in quantity over time. It is often encountered in various fields, including population dynamics, finance, and natural sciences. In the context of first-order recurrence relations, exponential growth can be succinctly represented by the recurrence relation:

$$a_{n+1} = r \cdot a_n$$

Here, (a_n) represents the quantity at time (n), and (r) is a constant called the growth rate. The term (a_{n+1}) is obtained by multiplying the previous value (a_n) by the growth rate (r).

To understand the behavior of this relation, let's consider an example. Suppose we have an initial quantity (a_0) and a growth rate (r) greater than 1 (indicating exponential growth). The sequence of values $(a_0, a_1, a_2, a_3, \ldots)$ will exhibit rapid growth, with each term significantly larger than the preceding one. The growth rate (r) determines how quickly the quantity increases and the larger (r) is, the faster the growth (Dumas, 2014).

The closed-form solution for exponential growth is given by:

$$a_n = a_0 \cdot r^n$$

This formula reveals that the quantity (a_n) grows exponentially with time, where (a_0) represents the initial value at time $(n = 0)$, and (r^n) represents the exponential factor. As (n) increases, the exponential term (r^n) becomes increasingly larger, leading to exponential growth (Figure 6.1).

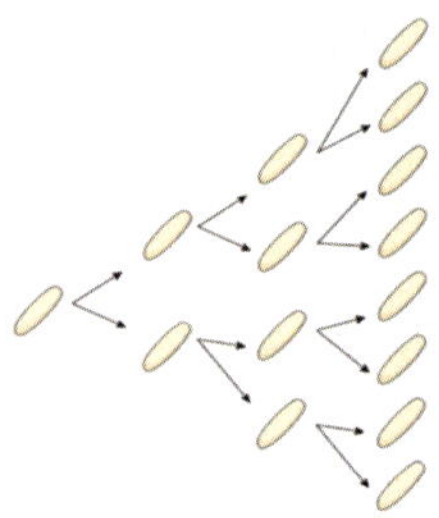

Figure 6.1. Illustration of an exponential growth.

Source: Nagwa, Creative Commons License.

6.2.2.2. Exponential Decay

Conversely, exponential decay models the continuous reduction of a quantity over time. It is observed in phenomena such as radioactive decay, the cooling of a heated object, or the decrease in the value of an investment. In the context of first-order recurrence relations, exponential decay can be described by the relation:

$$a_{n+1} = r \cdot a_n$$

In this case, (r) is a constant less than 1 (indicating decay), and each term (a_{n+1}) is obtained by multiplying the previous value (a_n) by the decay rate (r).

Consider an example where (a_0) represents the initial quantity, and (r) is a decay rate between 0 and 1. The sequence $(a_0, a_1, a_2, a_3, \ldots)$ will exhibit a continuous decrease, with each term being smaller than the preceding one (Bankole et al., 2017). The closer (r) is to 1, the slower the decay and the closer (r) is to 0, the faster the quantity diminishes (Figure 6.2).

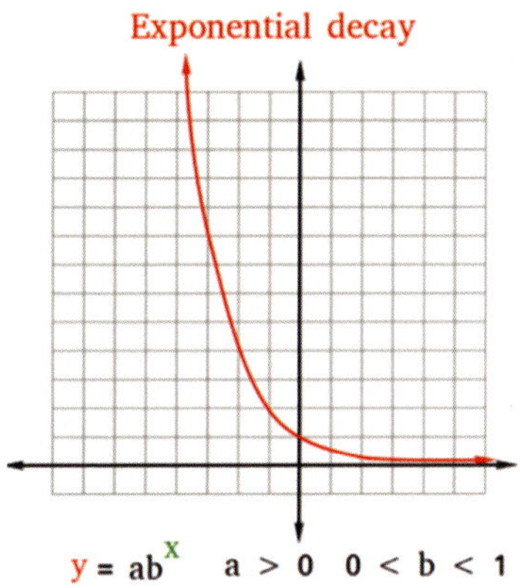

Figure 6.2. Example of an exponential decay.

Source: Basic Mathematics, Creative Commons License.

In summary, exponential growth and decay are fundamental concepts within first-order recurrence relations. These mathematical models provide valuable insights into the behavior of quantities that experience continuous change over time, whether through rapid increase (growth) or gradual decrease (decay). The growth or decay rate (r) plays a central role in determining the nature and speed of the change, making exponential growth and decay indispensable tools in various scientific and practical applications (Petkovšek, 1992).

6.2.3. Non-Homogeneous Relations

Non-homogeneous relations introduce external influences or forces that affect the sequence's progression beyond the inherent growth or decay described by the recurrence relation. In this section, we will explore non-homogeneous relations, their mathematical representation, and their implications in various mathematical and real-world contexts (Fiorenza & Vincenzi, 2011).

- **Mathematical Representation:** A non-homogeneous first-order recurrence relation can be formally expressed as:

$$a_{n+1} = r \cdot a_n + f(n)$$

Here, (a_{n+1}) represents the next term in the sequence, (a_n) is the current term, (r) is the recurrence coefficient or growth/decay rate, and $f(n)$ represents an additional function or term that influences the sequence's evolution at each step (n). This function $f(n)$ can take various forms and may depend on the discrete variable (n).

- **Interpretation and Examples:** Non-homogeneous recurrence relations arise in scenarios where external factors come into play, affecting the progression of a sequence. To understand this concept better, let's explore a few examples (Bagdasaryan, 2021).

1. **Linear Non-Homogeneous Relations:** Consider a scenario where a population undergoes exponential growth with a growth rate (r), but in addition to natural growth, there is a constant influx of new members. This situation can be modeled using a non-homogeneous relation:

$$a_{n+1} = r \cdot a_n + c$$

Here, (c) represents the constant rate of newcomers. This relation combines the inherent exponential growth with the constant addition of new members at each step.

2. **Nonlinear Non-Homogeneous Relations:** In some cases, the external influence on a sequence may not be represented by a simple linear term. For instance, in physics, the motion of an object under the influence of a non-constant force can be described by a non-homogeneous recurrence relation. The additional force function $f(n)$ would depend on the specific dynamics of the situation (Zhang & Ge, 2005).

3. **Real-World Applications:** Non-homogeneous recurrence relations find applications in various fields. For instance, in economics, a model for the value of an investment portfolio might consider both the exponential growth or decay of individual assets and external factors such as periodic contributions or withdrawals.
4. **Solving Non-Homogeneous Relations:** Solving non-homogeneous recurrence relations involves finding a closed-form expression for the sequence (a_n) that satisfies the given relation. This process typically combines solving the homogeneous part (i.e., the part without $f(n)$) and finding a particular solution that accounts for the non-homogeneous term $f(n)$. The general solution is then obtained by combining the homogeneous and particular solutions.

The specific approach to solving non-homogeneous relations may vary depending on the form of $f(n)$ and the recurrence relation itself. Methods such as variation of parameters, undetermined coefficients, and generating functions are often employed to find solutions (Bai & Su, 2005).

In conclusion, non-homogeneous recurrence relations encompass situations where external influences or forces impact the progression of a sequence described by a first-order recurrence relation. These relations are represented mathematically with an additional term $f(n)$ in the recurrence equation. Understanding and solving non-homogeneous relations are essential in various mathematical and real-world contexts, allowing us to model complex systems influenced by both inherent dynamics and external factors.

6.3. SECOND-ORDER RECURRENCE RELATIONS

In this section, we will study the second-order recurrence relations—an interesting area of mathematical study. These relations introduce an additional layer of complexity, describing sequences where each term depends not only on the previous term but also on the term before that. We will explore the mathematical foundations of second-order recurrence relations, including the crucial concepts of characteristic equations and roots. Furthermore, we will uncover their wide-ranging applications, from modeling oscillatory behavior to analyzing growth patterns. Real-world examples will illustrate the practical significance of these relations in diverse fields (Morita et al., 2018).

6.3.1. Characteristic Equations and Roots

In the study of second-order recurrence relations, one of the fundamental concepts is the use of characteristic equations and their associated roots. These mathematical tools play a pivotal role in understanding and solving second-order recurrence relations, which describe sequences where each term depends not only on the preceding term but also on the term before that. In this section, we delve into the intricacies of characteristic equations and roots within the context of second-order recurrence relations.

6.3.1.1. Mathematical Formulation

A second-order recurrence relation typically takes the form:

$$a_{n+2} = c_1 \cdot a_{n+1} + c_2 \cdot a_n$$

Here, (a_{n+2}) represents the next term in the sequence, (a_{n+1}) is the current term, and (a_n) is the term before that. The coefficients (c_1) and (c_2) are constants that determine how the sequence evolves from one step to the next. The task at hand is to find a solution to this recurrence relation, which involves determining the terms (a_n) for all (n) based on initial conditions (Maruotti & Rocci, 2012).

6.3.1.2. Characteristic Equation

To tackle second-order recurrence relations, we introduce a characteristic equation, which is derived from the relation's coefficients (c_1) and (c_2). The characteristic equation takes the form:

$$r^2 = c_1 \cdot r + c_2$$

In this equation, (r) represents the characteristic variable, and it is used to find the characteristic roots, denoted as (r_1) and (r_2), that satisfy the equation. These roots play a critical role in solving the recurrence relation and determining the sequence's behavior.

6.3.1.3. Characteristic Roots and Solutions

The characteristic roots (r_1) and (r_2) can be found by solving the characteristic equation. Depending on the values of (c_1) and (c_2), the roots can take different forms:

1. **Distinct Real Roots:** When the discriminant $(\ddot{A} = c_1^2 - 4c_2)$ is positive, the characteristic roots are distinct real numbers, denoted as (r_1) and (r_2). The general solution to the recurrence relation is:

$$a_n = A \cdot r_1^n + B \cdot r_2^n$$

Here, (A) and (B) are constants determined by the initial conditions.

2. **Complex Roots:** When the discriminant $(\ddot{A})$ is negative, the characteristic roots are complex conjugates of each other. The general solution involves complex exponentials:

$$a_n = e^{\alpha n}(C\cos(\beta n) + D\sin(\beta n))$$

Where $(\acute{a})$ and $(\hat{a})$ are real constants determined by the real and imaginary parts of the complex roots, and (C) and (D) are constants determined by the initial conditions.

3. **Repeated Real Roots:** In cases where the discriminant $(\ddot{A})$ is zero, the characteristic equation yields a repeated real root $(r_1 = r_2)$. The general solution is:

$$a_n = (A + Bn) \cdot r_1^n$$

Where (A) and (B) are constants determined by the initial conditions.

6.3.1.4. Significance and Applications

Characteristic equations and roots are crucial tools for solving second-order recurrence relations in various fields, including physics, engineering, and mathematics. They allow us to determine the general form of the sequence based on the initial conditions and the nature of the characteristic roots.

These solutions find applications in modeling oscillatory systems, analyzing growth patterns, and understanding the behavior of dynamic processes that involve second-order dependencies (Yang & Han, 2021).

In summary, characteristic equations and their associated roots are fundamental to the study of second-order recurrence relations. They provide a structured approach to solving these relations and unveil the underlying dynamics of sequences with dependencies on both the previous and penultimate terms.

Understanding the nature of the characteristic roots enables us to determine the general solution of the recurrence relation and gain insights into the behavior of the sequence it describes.

6.3.2. Applications in Oscillations and Growth

Second-order recurrence relations, characterized by their dependence on the two preceding terms in a sequence, find diverse applications in fields ranging from physics to engineering and mathematics. This section focuses on the specific applications of second-order recurrence relations in oscillatory phenomena and growth patterns. These mathematical tools allow us to model and analyze dynamic systems that exhibit periodic oscillations and various growth behaviors (Li et al., 2020).

6.3.2.1. Oscillations

Oscillations refer to the repetitive and periodic motion or behavior observed in numerous natural and engineered systems. Second-order recurrence relations play a pivotal role in understanding and modeling oscillations.

- **Simple Harmonic Oscillator:** A classic example of an oscillatory system is the simple harmonic oscillator, which describes the motion of a mass attached to a spring. The displacement of the mass at each moment depends on both its position and velocity. This relationship can be expressed using a second-order recurrence relation:

$$x_{n+2} = -\frac{k}{m} x_n$$

Here, (x_{n+2}) represents the position at time $(n+2)$, (x_n) is the position at time (n), (k) is the spring constant, and (m) is the mass of the object. Solving this recurrence relation reveals the sinusoidal nature of the oscillatory motion, with a period and frequency determined by the constants (k) and (m) (Figure 6.3).

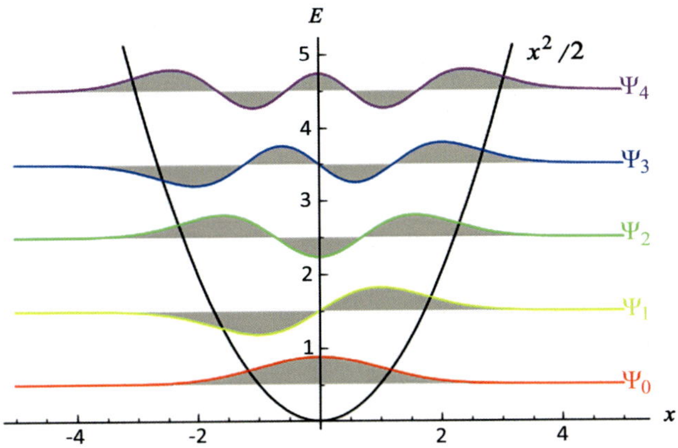

Figure 6.3. Illustration for the solution of simple harmonic motion equation.

Source: Mohammad Yasir, Creative Commons License.

6.3.2.2. Growth Patterns

Second-order recurrence relations also find applications in modeling various growth patterns, which can be observed in fields such as biology, finance, and population dynamics (Pan & Pan, 2016).

- **Population Growth:** In biology and ecology, second-order recurrence relations are used to model population growth and predator-prey interactions. For instance, the Lotka-Volterra equations describe the dynamics of predator and prey populations using a set of second-order recurrence relations. The solutions to these relations reveal the cyclical nature of population dynamics, with predator and prey populations oscillating over time.
- **Financial Modeling:** In finance, second-order recurrence relations are employed to analyze the growth of investments or assets. The recurrence relation may include factors such as interest rates, compounding periods, and initial investments. Solving these relations helps predict the future value of investments, taking into account periodic growth or compounding.

6.3.2.3. Real-World Examples

Let's consider a specific real-world example: the modeling of a simple pendulum's oscillatory motion. A simple pendulum consists of a mass (the bob) attached to a string or rod of fixed length. When displaced from its equilibrium position and released, the pendulum oscillates back and forth.

The motion of a simple pendulum can be described using a second-order recurrence relation:

$$\theta_{n+2} = -\frac{g}{L}\sin(\theta_n)$$

In this equation, (θ_{n+2}) represents the angular displacement of the pendulum at time $(n+2), (\theta_n)$ is the displacement at time (n), (g) is the acceleration due to gravity, and (L) is the length of the pendulum. Solving this recurrence relation reveals the periodic nature of the pendulum's oscillations, with the angle (θ) alternating between positive and negative values (Vetyukov et al., 2011).

6.4. HIGHER-ORDER RECURRENCE RELATIONS

Higher-order recurrence relations arise in various scientific and engineering disciplines, allowing us to model complex systems with a higher degree of precision. We will explore the fundamental concepts behind extending recurrence relations to higher orders, delve into the solutions involving complex roots, and examine practical examples that highlight the significance of these relations in solving real-world problems. By navigating through the intricacies of higher-order recurrence relations, we unlock new insights into the dynamic behavior of systems governed by these mathematical constructs.

6.4.1. Definition and Relation to Recurrence Relations

Higher-order recurrence relations represent a mathematical framework that extends the concept of recurrence relations to systems with more intricate dependencies. These relations encapsulate the idea that each term in a sequence is determined not only by its immediate predecessor but also by a specific combination of its preceding terms. By introducing higher degrees of dependence, these relations offer a powerful tool for modeling complex systems in various scientific and engineering domains (Verma, 1994).

Mathematically, a higher-order recurrence relation of order (k) can be expressed as:

$$a_{n+k} = F(a_n, a_{n+1}, \ldots, a_{n+k-1})$$

Here, (a_n) represents the nth term in the sequence, and the function (F) defines the relationship between the current term (a_{n+k}) and its (k) preceding terms $(a_n, a_{n+1}, \ldots, a_{n+k-1})$. The order (k) indicates the number of preceding terms influencing the current term. Higher-order recurrence relations encompass various special cases, including second-order (where $k = 2$) and third-order (where $k = 3$) relations, which have been discussed previously. These relations are integral to understanding dynamic systems with more complex dependencies and have applications in diverse fields, including physics, engineering, and mathematics.

6.4.2. Iterative Sequence Generation

One of the fundamental aspects of higher-order recurrence relations is their role in iterative sequence generation. Iteration implies the repetitive application of a specific rule or function to generate a sequence of values. In the context of higher-order recurrence relations, this rule is encapsulated within the function (F) that determines each term based on its (k) preceding terms (Yap, 2011). To illustrate this concept, let's consider a practical example involving a third-order recurrence relation. Suppose we have the following relation:

$$a_{n+3} = a_n + 2a_{n+1} - a_{n+2}$$

In this relation, each term (a_{n+3}) depends on the three preceding terms $(a_n), (a_{n+1})$, and (a_{n+2}). To generate a sequence, we start with initial values for $(a_0), (a_1)$, and (a_2) and then apply the recurrence relation iteratively. For instance, if we choose $(a_0 = 1), (a_1 = 2)$, and $(a_2 = 3)$ as our initial values, we can compute subsequent terms as follows:

$$a_3 = a_0 + 2a_1 - a_2 = 1 + 2(2) - 3 = 2$$

$$a_4 = a_1 + 2a_2 - a_3 = 2 + 2(3) - 2 = 8$$

$$a_5 = a_2 + 2a_3 - a_4 = 3 + 2(2) - 8 = -1$$

By repeating this process, we generate a sequence of values: $(1, 2, 3, 2, 8, -1, \ldots)$. This iterative sequence generation showcases the dynamic nature of higher-order recurrence relations and their ability to produce sequences with complex dependencies (Hitczenko et al., 2006).

6.4.3. Calculating Terms in Recursive Sequences

Calculating specific terms within a recursive sequence governed by higher-order recurrence relations requires a systematic approach that leverages the recurrence relation itself. The goal is to determine the value of a particular term in the sequence without explicitly computing all preceding terms.

Let's continue with the example of a third-order recurrence relation:

$$a_{n+3} = a_n + 2a_{n+1} - a_{n+2}$$

Suppose we wish to calculate (a_{10}) in this sequence. To do so, we can employ the recurrence relation iteratively, working our way up to the desired term:

1. We start with the initial values $(a_0),(a_1)$, and (a_2) and calculate (a_3) using the recurrence relation.
2. With $(a_0),(a_1),(a_2)$, and (a_3), we can compute (a_4).
3. This process continues until we reach a_{10}:

$$a_3 = a_0 + 2a_1 - a_2$$

$$a_4 = a_1 + 2a_2 - a_3$$

$$a_5 = a_2 + 2a_3 - a_4$$

$$a_6 = a_3 + 2a_4 - a_5$$

$$a_7 = a_4 + 2a_5 - a_6$$

$$a_8 = a_5 + 2a_6 - a_7$$

$$a_9 = a_6 + 2a_7 - a_8$$

$$a_{10} = a_7 + 2a_8 - a_9$$

By systematically applying the recurrence relation, we arrive at the value of (a_{10}). This approach allows us to calculate specific terms within recursive sequences efficiently, even in higher-order recurrence relations with complex dependencies.

Tip

When calculating terms in recursive sequences, it can be helpful to create a table or spreadsheet to organize your calculations and keep track of the values of each term.

In summary, higher-order recurrence relations extend the principles of recurrence relations to systems with more intricate dependencies, enabling us to model and analyze complex dynamic systems. These relations are integral to iterative sequence generation, and calculating terms within recursive sequences relies on the systematic application of the recurrence relation. Through a structured approach, we gain valuable insights into the behavior of sequences governed by higher-order recurrence relations (Bradshaw & Vignat, 2023).

6.5. MASTER THEOREM AND TIME COMPLEXITY

This theorem offers a systematic way to analyze the time complexity of recursive algorithms, a crucial aspect of computer science and computational mathematics. We will begin by discussing the essence of the Master Theorem and its application in determining the time complexity of recursive algorithms. Subsequently, we will study the broader landscape of time complexity analysis, highlighting the efficiency and performance of recursive algorithms (Olafsson & Pasquale, 2013).

6.5.1. The Master Theorem

The Master Theorem is a fundamental tool in the analysis of algorithmic time complexity, particularly for recursive algorithms. It provides a framework for determining the time complexity of divide-and-conquer algorithms, which often manifest in problems like sorting, searching, and graph traversal. The Master Theorem, originally formulated by Akra and Bazzi in the 1990s, serves as a powerful tool for classifying and solving recurrence relations that describe the time complexity of these algorithms (Figure 6.4).

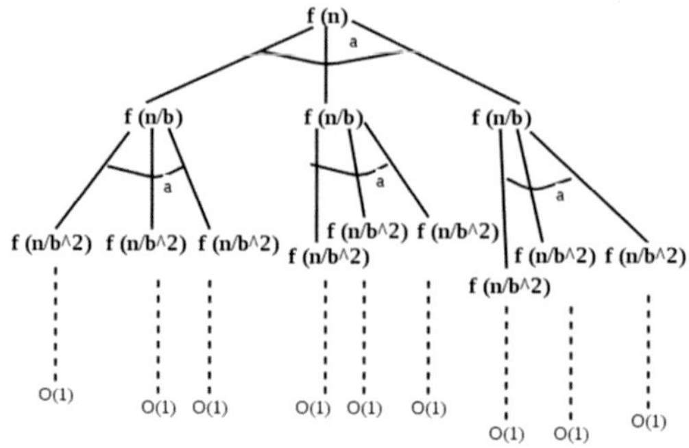

Figure 6.4. Master theorem in data structures.

Source: Geeks for Geeks, Creative Commons License.

Mathematically, the Master Theorem applies to recurrence relations of the form:

$$T(n) = aT\left(\frac{n}{b}\right) + f(n)$$

Here, $T(n)$ represents the time complexity of the algorithm for a problem of size (n). The recurrence relation divides the problem into (a) subproblems of size (n/b) and combines their solutions, with $f(n)$ denoting the time spent on work outside of the recursive subproblems (Soares et al., 2022).

To utilize the Master Theorem, one must identify the values of (a), (b), and the form of $f(n)$ to determine the time complexity class of the algorithm.

The Master Theorem classifies recurrence relations into three main cases:

- **Case 1:**

If $f(n)$ is $O\left(n^{\log_b a-\epsilon}\right)$ for some $(\epsilon > 0)$, where $(\log_b a)$ is the exponent of the recursive part, then the time complexity is $T(n) = \Theta\left(n^{\log_b a}\right)$.

In this case, the work done outside the recursive subproblems is subdominant, and the time complexity is determined primarily by the recursive part. This scenario often occurs in divide-and-conquer algorithms, where each subproblem contributes significantly to the overall time (Mansour, 2004).

- **Case 2:**

If $f(n)$ is $\Theta(n^{\log_b a})$, where $(\log_b a)$ is the exponent of the recursive part,

then the time complexity is $T(n) = \Theta(n^{\log_b a} \log n)$.

Here, the work done outside the recursive subproblems is precisely balanced with the recursive part. The logarithmic factor $(\log n)$ accounts for the extra work needed to combine the subproblem solutions.

- **Case 3:**

If $f(n)$ is $\grave{U}(n^{\log_b a+\epsilon})$ for some $(\epsilon > 0)$, and if $af(n/b) \leq kf(n)$ for some $(k < 1)$ and sufficiently large (n), then the time complexity is $T(n) = \Theta(f(n))$.

In this case, the work done outside the recursive subproblems dominates, and the time complexity is primarily determined by $f(n)$. This situation often arises when the extra work needed to combine subproblem solutions outweighs the recursive work.

6.5.2. Time Complexity Analysis

Time complexity analysis is a crucial aspect of algorithm design and analysis. It provides a quantitative measure of an algorithm's efficiency by estimating the computational resources required as a function of the input size. Understanding the time complexity of an algorithm helps in making informed decisions about algorithm selection and optimization.

Did You Know?

Time complexity analysis, often related to recurrence relations, is crucial in computer science and algorithm design for understanding the efficiency of algorithms.

The time complexity of an algorithm is typically expressed using big-O notation, which provides an upper bound on the growth rate of the algorithm's running time as a function of the input size. In the context of recursive algorithms, time complexity analysis involves solving recurrence relations to determine how the algorithm's running time scales with the input size (Schoenberg, 1955).

6.5.3. Recursive Algorithms and Efficiency

Recursive algorithms are a common paradigm in computer science and mathematics, particularly in divide-and-conquer approaches. They offer elegant solutions to complex problems by breaking them down into smaller, more manageable subproblems. However, the efficiency of recursive algorithms depends on various factors, including the problem structure, the choice of recursion strategy, and the time complexity.

Efficiency in the context of recursive algorithms encompasses several key considerations:

1. **Time Complexity:** The time complexity of a recursive algorithm quantifies the computational resources required as a function of the input size. Efficient recursive algorithms aim to minimize time complexity, often by optimizing the recurrence relations involved.
2. **Space Complexity:** Recursive algorithms may use additional memory to store recursive calls and intermediate results on the call stack. Evaluating space complexity is crucial to avoid excessive memory usage and potential stack overflow errors.
3. **Tail Recursion:** Tail recursion is a specific form of recursion where the recursive call is the last operation in a function. Tail-recursive algorithms can be optimized by compilers or interpreters to eliminate unnecessary function calls, reducing both time and space complexity.
4. **Memoization and Dynamic Programming**: In some cases, recursive algorithms can benefit from memoization or dynamic programming techniques to store and reuse intermediate results. These strategies can significantly improve efficiency by avoiding redundant computations.
5. **Choosing the Right Data Structures:** Recursive algorithms often involve data structures like trees or graphs. Efficient data structure selection is crucial for minimizing time and space complexity. Balancing the trade-offs between space and time complexity is essential.

6. **Optimizing the Base Case:** The base case of a recursive algorithm is critical, as it determines when the recursion terminates. Optimizing the base case can lead to significant efficiency improvements.

Overall, the efficiency of recursive algorithms depends on a combination of algorithm design, careful analysis of time and space complexity, and optimization strategies. By considering these factors and understanding the time complexity using tools like the Master Theorem, one can design and implement recursive algorithms that strike a balance between elegance and computational efficiency (Fu & Chang, 2002).

6.6. ADVANCED TOPICS IN RECURRENCE RELATIONS

In this section, we will study some advanced topics in recurrence relations, exploring more complex and intricate aspects of these mathematical constructs. These topics are of great significance in various mathematical and scientific disciplines, offering deeper insights into the behavior of dynamic systems and providing powerful tools for problem-solving.

6.6.1. Nonlinear Relations and Solutions

While many introductory recurrence relations are linear in nature, advanced applications often involve nonlinear recurrence relations, which can exhibit more intricate and nuanced behaviors. These relations can give rise to complex and chaotic dynamics, making them essential in fields like physics, biology, and economics (Shapiro et al., 2022). Nonlinear recurrence relations may take various forms, such as exponential growth or decay, logistic maps, and chaotic systems. Solving nonlinear recurrence relations typically requires specialized mathematical techniques, including numerical methods and iterative approaches. Understanding the solutions to nonlinear recurrence relations is crucial for modeling and predicting the behavior of systems in which linear approximations are inadequate.

6.6.2. Generating Functions

Generating functions are a powerful mathematical tool for analyzing and solving recurrence relations, particularly in combinatorics and probability theory. A generating function is a formal power series that encodes the sequence of coefficients of a sequence in a compact and manipulable form.

These functions provide a bridge between the discrete sequences and the continuous functions, offering a unified framework for solving a wide range of recurrence relations (Figure 6.5).

Figure 6.5. Use of a binary tree for counting generating functions.

Source: Robert Sedgewick, Creative Commons License.

Generating functions come in various forms, including ordinary generating functions and exponential generating functions, each suited to specific types of problems. They facilitate the extraction of information about sequences, such as counting, summation, and asymptotic behavior, by transforming recurrence relations into algebraic equations that are easier to solve. Generating functions find applications in combinatorial enumeration, probability distributions, and the analysis of algorithms.

6.6.3. Advanced Methods and Applications

The realm of recurrence relations extends beyond the classic cases encountered in introductory courses. Advanced methods and applications encompass a diverse array of topics and techniques used to solve complex recurrence relations and address intricate problems.

1. **Divide-and-Conquer Algorithms:** Advanced divide-and-conquer algorithms often involve intricate recurrence relations

that require specialized analysis. Examples include the Karatsuba algorithm for fast multiplication and algorithms for solving linear systems.

2. **Analytic Combinatorics:** Analytic combinatorics employs tools from complex analysis to analyze the behavior of combinatorial structures and their generating functions. This approach allows for precise asymptotic analysis of sequences.
3. **Stochastic Recurrence Relations:** In fields like stochastic processes and finance, recurrence relations involving randomness play a crucial role. Advanced methods involve probability theory and statistical analysis to understand the behavior of such systems (Kim et al., 2008).
4. **Chaos and Fractals:** Nonlinear recurrence relations can give rise to chaotic and fractal behavior. Understanding and characterizing the dynamics of chaotic systems involve advanced mathematical techniques and concepts from dynamical systems theory.
5. **Number Theory:** Recurrence relations are prevalent in number theory, particularly in the study of integer sequences and Diophantine equations. Advanced methods in this context often involve algebraic number theory and modular forms.
6. **Optimization and Control:** In engineering and control theory, recurrence relations are used to model dynamic systems and optimize control strategies. Advanced applications involve solving complex optimization problems with recurrence constraints.
7. **Parallel and Distributed Computing:** In computer science, recurrence relations play a role in the analysis of parallel and distributed algorithms. Advanced techniques address issues related to load balancing and parallelism.

These advanced methods and applications underscore the versatility and significance of recurrence relations across various scientific disciplines. They provide powerful tools for understanding complex systems, modeling dynamic behavior, and solving intricate problems that transcend the scope of elementary recurrence relations. Mastery of these advanced topics equips mathematicians, scientists, and engineers with the tools needed to tackle complex real-world challenges (Held & Karp, 1962).

Activity 6.1

Consider a scenario where you are designing a recursive algorithm to solve a complex computational problem. Describe the steps you would take to analyze the time complexity of your algorithm using recurrence relations and the Master Theorem. How would you optimize your algorithm based on this analysis?

6.7. SUMMARY

This chapter delves into the fundamental concepts of recurrence relations, providing a comprehensive exploration of their various aspects. Beginning with an introduction to recurrence relations, the chapter progresses to cover first-order recurrence relations, their solutions in combinatorics, and applications in exponential growth and decay. Second-order recurrence relations are discussed in detail, emphasizing characteristic equations and their relevance in understanding oscillations and growth patterns. The chapter further explores higher-order recurrence relations, iterative sequence generation, and time complexity analysis using the Master Theorem, a critical tool in computer science. Advanced topics, including nonlinear relations, generating functions, and their applications, are also presented, offering a deep understanding of solving complex recurrence relations. This chapter equips readers with valuable knowledge applicable to mathematics, computer science, and algorithm design.

REFERENCES

1. Abramovich, S., & Pieper, A., (1995). Fostering recursive thinking in combinatorics through the use of manipulatives and computing technology. *The Mathematics Educator, 6*(2), 364–366.
2. Albert, E., Arenas, P., Genaim, S., & Puebla, G., (2008). Automatic inference of upper bounds for recurrence relations in cost analysis. In: *Static Analysis: 15th International Symposium, SAS 2008, Valencia, Spain, July 16–18, 2008; Proceedings 15* (Vol. 1, pp. 221–237). Springer Berlin Heidelberg.
3. Ali, K. M., & Khan, M., (2019). Application based construction and optimization of substitution boxes over 2D mixed chaotic maps. *International Journal of Theoretical Physics, 58*, 3091–3117.
4. Ash, C. J., (1986). Recursive labelling systems and stability of recursive structures in hyperarithmetical degrees. *Transactions of the American Mathematical Society, 298*(2), 497–514.
5. Bagdasaryan, A. G., (2021). Solving third-order linear recurrence relations with applications to number theory and combinatorics. In: *International Conference on Difference Equations and Applications* (Vol. 2, pp. 97–121). Cham: Springer International Publishing.
6. Bai, Z., & Su, Y., (2005). SOAR: A second-order Arnoldi method for the solution of the quadratic eigenvalue problem. *SIAM Journal on Matrix Analysis and Applications, 26*(3), 640–659.
7. Bankole, P. A., Ojo, E. K., & Odumosu, M. O., (2017). On recurrence relations and application in predicting price dynamics in the presence of economic recession. *International Journal of Discrete Mathematics, 2*(4), 125–131.
8. Berthomieu, J., & Faugère, J. C., (2016). Guessing linear recurrence relations of sequence tuples and p-recursive sequences with linear algebra. In: *Proceedings of the ACM on International Symposium on Symbolic and Algebraic Computation* (Vol. 1, pp. 95–102).
9. Bradshaw, Z. P., & Vignat, C., (2023). An operational calculus generalization of Ramanujan's master theorem. *Journal of Mathematical Analysis and Applications, 523*(2), 127029.
10. Danoun, A., Prulière, E., & Chemisky, Y., (2022). Thermodynamically consistent recurrent neural networks to predict nonlinear behaviors of dissipative materials subjected to non-proportional loading paths. *Mechanics of Materials, 173*, 104436.

11. Dumas, P., (2014). Asymptotic expansions for linear homogeneous divide-and-conquer recurrences: Algebraic and analytic approaches collated. *Theoretical Computer Science, 548*, 25–53.
12. Fiorenza, A., & Vincenzi, G., (2011). Limit of ratio of consecutive terms for general order-k linear homogeneous recurrences with constant coefficients. *Chaos, Solitons & Fractals, 44*(1–3), 145–152.
13. Fiorini, R., (2015). Acybernetics update for competitive deep learning system. In: *Proceedings of the 2nd International Conference on Entropy and Its Applications* (Vol. 1, pp. 1–27). MDPI.
14. Fu, J. C., & Chang, Y. M., (2002). On probability generating functions for waiting time distributions of compound patterns in a sequence of multistate trials. *Journal of Applied Probability, 39*(1), 70–80.
15. Gautschi, W., (1967). Computational aspects of three-term recurrence relations. *SIAM Review, 9*(1), 24–82.
16. Held, M., & Karp, R. M., (1962). A dynamic programming approach to sequencing problems. *Journal of the Society for Industrial and Applied Mathematics, 10*(1), 196–210.
17. Hitczenko, P., Johnson, J. R., & Huang, H. J., (2006). Distribution of a class of divide and conquer recurrences arising from the computation of the Walsh–Hadamard transform. *Theoretical Computer Science, 352*(1–3), 8–30.
18. Hollerbach, J. M., (1980). A recursive Lagrangian formulation of manipulator dynamics and a comparative study of dynamics formulation complexity. *IEEE Transactions on Systems, Man, and Cybernetics, 10*(11), 730–736.
19. Hongwei, L. I., Yunlong, Z. H. O. U., Bin, S. U. N., & Yue, Y. A. N. G., (2010). Multi-scale chaotic analysis of the characteristics of gas-liquid two-phase flow patterns. *Chinese Journal of Chemical Engineering, 18*(5), 880–888.
20. Hsu, C. P., & Marcus, R. A., (1997). A sequential formula for electronic coupling in long range bridge-assisted electron transfer: Formulation of theory and application to alkanethiol monolayers. *The Journal of Chemical Physics, 106*(2), 584–598.
21. Hupkes, D., Veldhoen, S., & Zuidema, W., (2018). Visualization and 'diagnostic classifiers' reveal how recurrent and recursive neural networks process hierarchical structure. *Journal of Artificial*

Intelligence Research, 61, 907–926.

22. Kara, E. E., & Ilkhan, M., (2016). Some properties of generalized Fibonacci sequence spaces. *Linear and Multilinear Algebra, 64*(11), 2208–2223.
23. Kessler, C. M., & Anderson, J. R., (1986). Learning flow of control: Recursive and iterative procedures. *Human-Computer Interaction, 2*(2), 135–166.
24. Kim, S., Gupta, N., & Pevzner, P. A., (2008). Spectral probabilities and generating functions of tandem mass spectra: A strike against decoy databases. *Journal of Proteome Research, 7*(8), 3354–3363.
25. Li, Y., Dong, Y., & Qian, J., (2020). Higher-order analysis of probabilistic long-term loss under nonstationary hazards. *Reliability Engineering & System Safety, 203*, 107092.
26. Liu, Y. A., & Stoller, S. D., (1999). From recursion to iteration: What are the optimizations? In: *Proceedings of the 2000 ACM SIGPLAN Workshop on Partial Evaluation and Semantics-Based Program Manipulation* (Vol. 2, pp. 73–82).
27. Mansour, T., (2004). A formula for the generating functions of powers of Horadam's sequence. *Australasian Journal of Combinatorics, 30*, 207–212.
28. Martins, M. D., Martins, I. P., & Fitch, W. T., (2016). A novel approach to investigate recursion and iteration in visual hierarchical processing. *Behavior Research Methods, 48*, 1421–1442.
29. Maruotti, A., & Rocci, R., (2012). A mixed non-homogeneous hidden Markov model for categorical data, with application to alcohol consumption. *Statistics in Medicine, 31*(9), 871–886.
30. Morita, L. H. M., Tomazella, V. L., & Louzada-Neto, F., (2018). Accelerated lifetime modelling with frailty in a non-homogeneous Poisson process for analysis of recurrent events data. *Quality Technology & Quantitative Management, 15*(2), 209–229.
31. Necula, G. C., (2000). Translation validation for an optimizing compiler. In: *Proceedings of the ACM SIGPLAN 2000 Conference on Programming Language Design and Implementation* (Vol. 2, pp. 83–94).
32. Olafsson, G., & Pasquale, A., (2013). Ramanujan's master theorem for the hypergeometric Fourier transform associated with root systems. *Journal of Fourier Analysis and Applications, 19*, 1150–1183.

33. Pan, G., Ou, R., & Faulwasser, T., (2023). Data-driven stochastic output-feedback predictive control: Recursive feasibility through interpolated initial conditions. In: *Learning for Dynamics and Control Conference* (Vol. 1, pp. 980–992). PMLR.
34. Pan, S. W., & Pan, J. Q., (2016). Direct solutions of linear non-homogeneous difference equations. *Advances in Difference Equations, 2016*(1), 1–10.
35. Petkovšek, M., (1992). Hypergeometric solutions of linear recurrences with polynomial coefficients. *Journal of Symbolic Computation, 14*(2–3), 243–264.
36. Poland, D., (1974). Recursion relation generation of probability profiles for specific-sequence macromolecules with long-range correlations. *Biopolymers: Original Research on Biomolecules, 13*(9), 1859–1871.
37. Protter, M., Yavneh, I., & Elad, M., (2010). Closed-form MMSE estimation for signal denoising under sparse representation modeling over a unitary dictionary. *IEEE Transactions on Signal Processing, 58*(7), 3471–3484.
38. Renault, M., (2013). The period, rank, and order of the (a, b)-Fibonacci sequence mod m. *Mathematics Magazine, 86*(5), 372–380.
39. Schoenberg, I. J., (1955). On the zeros of the generating functions of multiply positive sequences and functions. *Annals of Mathematics, 62*(3), 447–471.
40. Shannon, A. G., (1974). Some properties of a fundamental recursive sequence of arbitrary order. *The Fibonacci Quarterly, 12*(4), 327–335.
41. Shapiro, L., Sprugnoli, R., Barry, P., Cheon, G. S., He, T. X., Merlini, D., & Wang, W., (2022). Extraction of coefficients and generating functions. In: *The Riordan Group and Applications* (Vol. 2, pp. 19–46). Cham: Springer International Publishing.
42. Smith, B. E., (2017). Composing across modes: A comparative analysis of adolescents' multimodal composing processes. *Learning, Media, and Technology, 42*(3), 259–278.
43. Soares, A., Street, A., Andrade, T., & Garcia, J. D., (2022). An integrated progressive hedging and benders decomposition with multiple master method to solve the Brazilian generation expansion problem. *IEEE Transactions on Power Systems, 37*(5), 4017–4027.
44. Stoll, M., (1997). Bounds for the length of recurrence relations for convolutions of P-recursive sequences. *European Journal of Combinatorics, 18*(6), 707–712.

45. Thurston, M., (1999). "A deliberate collection of cross purposes": Eavan Boland's poetic sequences. *Colby Quarterly, 35*(4), 5.
46. Torres, F., Torres, S. N., & Martín, C. S., (2005). A recursive least square adaptive filter for nonuniformity correction of infrared image sequences. In: *Progress in Pattern Recognition, Image Analysis and Applications: 10th Iberoamerican Congress on Pattern Recognition, CIARP 2005, Havana, Cuba, November 15–18, 2005; Proceedings 10* (Vol. 1, pp. 540–546). Springer Berlin Heidelberg.
47. Treacy, M. M. J., Newsam, J. M., & Deem, M. W., (1991). A general recursion method for calculating diffracted intensities from crystals containing planar faults. *Proceedings of the Royal Society of London. Series A: Mathematical and Physical Sciences, 433*(1889), 499–520.
48. Verma, R. M., (1994). A general method and a master theorem for divide-and-conquer recurrences with applications. *Journal of Algorithms, 16*(1), 67–79.
49. Vetyukov, Y., Kuzin, A., & Krommer, M., (2011). Asymptotic splitting in the three-dimensional problem of elasticity for non-homogeneous piezoelectric plates. *International Journal of Solids and Structures, 48*(1), 12–23.
50. Viswanath, D., (2000). Random Fibonacci sequences and the number 1.13198824.... *Mathematics of Computation, 69*(231), 1131–1155.
51. Yang, B. C., & Han, X. F., (2021). A new method for solving the linear recurrence relation with nonhomogeneous constant coefficients in some cases. In: *Journal of Physics: Conference Series* (Vol. 2113, pp. 120–170). IOP Publishing.
52. Yap, C., (2011). A real elementary approach to the master recurrence and generalizations. In: *International Conference on Theory and Applications of Models of Computation* (Vol. 2, pp. 14–26). Berlin, Heidelberg: Springer Berlin Heidelberg.
53. Zhang, Y., & Ge, S. S., (2005). Design and analysis of a general recurrent neural network model for time-varying matrix inversion. *IEEE Transactions on Neural Networks, 16*(6), 1477–1490.
54. Zhou, J., Zhu, Y., Li, X. R., & You, Z., (2002). Variants of the Greville formula with applications to exact recursive least squares. *SIAM Journal on Matrix Analysis and Applications, 24*(1), 150–164.

CHAPTER 7

Lucas Number and Higher-Order Sequence

Contents

7.1. INTRODUCTION

In this chapter, Lucas numbers will be discussed in detail, which is a sequence of integers with various mathematical properties and applications. Lucas numbers, denoted as form a sequence that shares similarities with the more well-known Fibonacci numbers. This chapter begins with the fundamental definition and properties of Lucas numbers, followed by an investigation into their intriguing relationship with Fibonacci numbers. Additionally, we delve into various formulae and identities that characterize the behavior of Lucas numbers, highlighting their mathematical significance (Ramirez, 2015) (Figure 7.1).

Lucas Numbers	
1st	1
2nd	3
3rd	4
4th	7
5th	11
6th	18
7th	29
8th	47
9th	76
10th	123
11th	199
12th	322
13th	521
14th	843
15th	1364

Figure 7.1. Schematic of Lucas numbers.

Source: Find the Factors, Creative Commons License.

7.1.1. Definition and Properties

Lucas numbers are an integer sequence defined recursively, just like the Fibonacci numbers. They are named after the French mathematician Édouard Lucas, who extensively studied their properties in the 19th century. The definition of Lucas numbers starts with two initial values:

$$L_0 = 2, \quad L_1 = 1$$

Subsequent terms are generated using the following recurrence relation:

$$L_{n+2} = L_{n+1} + L_n$$

This recurrence relation implies that each Lucas number is the sum of its two predecessors, mirroring the defining characteristic of the Fibonacci sequence.

Lucas numbers exhibit several interesting properties:

1. **Additive Property**: The fundamental recurrence relation of Lucas numbers reflects their additive nature. Each term is the sum of the two preceding terms.
2. **Initial Values:** The initial values $(L_0 = 2)$ and $(L_1 = 1)$ are crucial in generating the entire sequence. These values set the foundation for all subsequent Lucas numbers.
3. **Positive Integers:** Lucas numbers are a sequence of positive integers. As (n) increases, $(\quad)$ remains a positive whole number.
4. **Growth Rate:** The growth rate of Lucas numbers is approximately the golden ratio, denoted as (f), which is equal to $\left(\frac{1+\sqrt{5}}{2}\right)$. This is a remarkable connection shared with the Fibonacci sequence.

7.1.2. Relationship with Fibonacci Numbers

One of the most captivating aspects of Lucas numbers is their intimate relationship with the Fibonacci numbers. This relationship can be succinctly expressed as follows:

$$L_n = F_{n-1} + F_{n+1}$$

Here, (F_n) represents the nth Fibonacci number. The connection between Lucas and Fibonacci numbers becomes evident when we realize that each Lucas number is the sum of the Fibonacci number immediately preceding it (F_{n-1}) and the Fibonacci number immediately following it (F_{n+1}).

This relationship highlights the intertwined nature of these two sequences, with each sequence influencing the other. It also underscores the significant role played by the golden ratio (f) in the behavior of both Lucas and Fibonacci numbers (Koshy, 2019) (Figure 7.2).

Figure 7.2. A program for Fibonacci numbers.

Geeks for Geeks, Creative Commons License.

7.1.3. Formulae and Identities

Lucas numbers are not just defined by their recurrence relation; they can also be expressed in various closed-form formulae and exhibit fascinating mathematical identities.

Explicit Formula: The nth Lucas number can be explicitly calculated using the formula:

$$L_n = f^n + (-f)^{-n}$$

Here, (f) is the golden ratio.

- **Binet-Lucas Formula:**

The Binet-Lucas formula, analogous to the Binet formula for Fibonacci numbers, allows for the direct calculation of Lucas numbers using (f) and its conjugate $(\hat{\phi})$:

$$L_n = \frac{\phi^n - \widehat{\phi^n}}{\sqrt{5}}$$

($\hat{\Phi}$) is the conjugate of (f) and is approximately equal to -0.6180339887.

- **Cassini's Identity:**

Lucas numbers satisfy Cassini's identity, which states that for all positive integers (n):

$$L_{n-1} \cdot L_{n+1} - L_n^2 = (-1)^n$$

- **Reciprocal Identity:**

Lucas numbers also exhibit a reciprocal identity, connecting (L_n) and (L_{n-1}):

$$\frac{1}{L_n} + \frac{1}{L_{n-1}} = \frac{5}{L_{n-1} \cdot L_n}$$

These formulae and identities provide valuable tools for calculating Lucas numbers, establishing relationships between them, and exploring their mathematical properties.

We have laid the foundation for understanding Lucas numbers by defining them, exploring their key properties, and uncovering their intricate relationship with Fibonacci numbers.

Additionally, we have introduced essential formulae and identities that will serve as valuable tools in our further exploration of Lucas numbers and their applications (Falcon, 2011).

7.2. GENERALIZED LUCAS SEQUENCES

While Lucas numbers, denoted as (L_n), are the most well-known. They represent just the tip of the iceberg. Generalized Lucas sequences encompass a broader class of sequences, each with its unique properties and characteristics.

We begin by examining how Lucas sequences can be extended to higher orders, followed by a detailed exploration of their recursive definitions and initial conditions. To illustrate the concepts, we provide concrete examples that showcase the diversity and mathematical richness of these sequences.

7.2.1. Extending to Higher Orders

Lucas sequences, at their core, are defined by a second-order linear recurrence relation. However, the concept of generalized Lucas sequences allows us to extend these sequences to higher orders, opening up a world of possibilities in the integer sequences. A generalized Lucas sequence of order (k) is defined by a recurrence relation of the form:

$$L_{n+k} = a_1 L_{n+k-1} + a_2 L_{n+k-2} + \ldots + a_k L_n$$

Here, $(a_1, a_2, \ldots, a_k)$ are coefficients that determine the sequence's behavior. This higher-order recurrence relation builds upon the fundamental principles of Lucas sequences while introducing additional complexity and diversity (Lengyel, 1995) (Figure 7.3).

Example of a Lucas Sequence

$P = 2$ and $Q = -1$

$U_0 = 0$	$V_0 = 2$
$U_1 = 1$	$V_1 = 2$
$U_2 = 2$	$V_2 = 6$
$U_3 = 5$	$V_3 = 14$
$U_4 = 12$	$V_4 = 34$
$U_5 = 29$	$V_5 = 82$
$U_6 = 70$	$V_6 = 198$
$U_7 = 169$	$V_7 = 478$
$U_8 = 408$	$V_8 = 1154$

Figure 7.3. Example of Lucas sequence.

Source: Cathleen Craig, Creative Commons License.

Generalized Lucas sequences of higher orders are a significant field of study with connections to various areas of mathematics. They offer unique challenges and opportunities for exploration, making them a captivating subject for mathematicians and researchers.

7.2.2. Recursive Definitions and Initial Conditions

Understanding generalized Lucas sequences necessitates a thorough grasp of their recursive definitions and initial conditions. The recursive definition of a generalized Lucas sequence of order (k) provides a formula for each term (L_n) in terms of its predecessors $(L_{n-1}, L_{n-2}, \ldots, L_{n-k})$. The coefficients

$(a_1, a_2, \ldots, a_k)$ dictate the specific form of this recurrence relation (Koken & Bozkurt, 2008).

Additionally, the initial conditions are critical for uniquely defining a generalized Lucas sequence. These initial values, denoted as $(L_0, L_1, \ldots, L_{k-1})$, establish the starting point for the sequence. The choice of initial conditions significantly impacts the entire sequence's behavior, making it a pivotal aspect of sequence theory.

The combination of the recursive definition and initial conditions allows us to construct and explore the properties of generalized Lucas sequences. It is through this mathematical framework that we can unravel the sequence's intricacies and uncover its hidden patterns.

To illustrate the concept of generalized Lucas sequences and their diversity, let's delve into some concrete examples. We will explore sequences of various orders, each with its unique characteristics and properties (Dresden & Du, 2014).

- **Example: Second-Order Lucas Sequence** $(k \quad 2)$

Let's consider a generalized Lucas sequence of order 2 with coefficients $(a_1 = 1)$ and $(a_2 = 2)$. The initial conditions are $(L_0 = 1)$ and $(L_1 = 3)$. Using the recurrence relation, we can generate the sequence as follows:

$$L_2 = a_1 L_1 + a_2 L_0 = 1 \cdot 3 + 2 \cdot 1 = 5$$

$$L_3 = a_1 L_2 + a_2 L_1 = 1 \cdot 5 + 2 \cdot 3 = 11$$

$$L_4 = a_1 L_3 + a_2 L_2 = 1 \cdot 11 + 2 \cdot 5 = 21$$

This sequence continues to grow, with each term depending on the two preceding terms. It exhibits its own unique properties and characteristics, distinct from the traditional Lucas numbers.

- **Example: Third-Order Lucas Sequence** $(k = 3)$

Now, let's explore a generalized Lucas sequence of order 3 with coefficients $(a_1 = 1), (a_2 = 2)$, and $(a_3 = -1)$. The initial conditions are $(L_0 = 2), (L_1 = 1)$, and $(L_2 = 4)$. Using the recurrence relation, we can generate the sequence:

$$L_3 = a_1L_2 + a_2L_1 + a_3L_0 = 1\cdot4 + 2\cdot1 - 1\cdot2 = 4$$

$$L_4 = a_1L_3 + a_2L_2 + a_3L_1 = 1\cdot4 + 2\cdot4 - 1\cdot1 = 11$$

$$L_5 = a_1L_4 + a_2L_3 + a_3L_2 = 1\cdot11 + 2\cdot4 - 1\cdot4 = 15$$

This third-order Lucas sequence demonstrates a different set of properties and behavior compared to the second-order example (Kilic, 2008).

These examples showcase the versatility and mathematical richness of generalized Lucas sequences, emphasizing their dependence on the order, coefficients, and initial conditions. The exploration of these sequences extends beyond these introductory examples, offering mathematicians a vast landscape to investigate and uncover new mathematical phenomena.

7.3. LUCAS NUMBERS IN NATURE AND MATHEMATICS

In this section, we explore the intriguing relationship between Lucas numbers and the natural world, as well as their profound connections to prime numbers and their role in algebraic number theory (Razpet, 2002).

7.3.1. Occurrence in Natural Phenomena

Nature often unfolds its mysteries through mathematical patterns and sequences, and Lucas numbers are no exception. Their occurrence in various natural phenomena underscores the ubiquity of mathematical patterns in the world around us.

7.3.1.1. Phyllotaxis and Botanical Arrangements

In phyllotaxis, the arrangement of leaves, seeds, or petals in plants, the number of spirals often corresponds to consecutive Lucas numbers. This phenomenon can be mathematically expressed as follows:

$$L_n = L_{n-1} + L_{n-2}$$

For example, consider a sunflower with 21 spirals in one direction and 34 spirals in the opposite direction. These numbers correspond to Lucas numbers (L_8) and (L_9), respectively. The recurrence relation $(L_n = L_{n-1} + L_{n-2})$ mirrors the spirals' growth, demonstrating the mathematical regularity in botanical structures (Ballabio et al., 2016).

7.3.1.2. Bee Populations

In bee populations, Lucas numbers can be observed in the organization of bee colonies. Bees tend to arrange themselves in patterns governed by Lucas numbers. For instance, consider a beehive where each cell contains 5 bees, and the next cell contains 8 bees. These numbers correspond to Lucas numbers (L_5) and (L_6), respectively. The recurrence relation captures the mathematical order in bee colony arrangements (Szynal, 2018).

7.3.1.3. Pinecone Spirals

Pinecones exhibit spiral patterns, and the number of spirals often corresponds to Lucas numbers. This phenomenon can be expressed mathematically as:

$$L_n = L_{n-1} + L_{n-2}$$

For instance, if a pinecone has 3 spirals in one direction and 5 spirals in the opposite direction, these numbers correspond to Lucas numbers (L_3) and (L_4).

These mathematical relationships illustrate the presence of Lucas numbers in natural phenomena and underscore the role of mathematical patterns in shaping the structure and organization of living organisms (Kilic & Tasci, 2006).

7.3.2. Connection to Prime Numbers

Lucas numbers exhibit a captivating connection to prime numbers, adding complexity to the study of both prime numbers and Lucas sequences.

Tip

Remember that Lucas numbers, like Fibonacci numbers, have intriguing connections to prime numbers. Explore these connections to deepen your understanding of both prime numbers and Lucas sequences.

7.3.2.1. Lucas-Lehmer Test for Mersenne Primes

The Lucas-Lehmer test is a specialized primality test used to determine whether a Mersenne number is prime. Mersenne numbers are of the form

$(2^n - 1)$, where (n) is a prime number. To test the primality of a Mersenne number, the Lucas-Lehmer test relies on Lucas numbers. Specifically, it checks whether a particular Lucas sequence generated by Lucas numbers is divisible by the Mersenne number in question. This can be expressed mathematically as follows:

$$L_{n+2} = (P-1)L_n - PL_{n-2}$$

Where $(P = 2^n)$ is the Mersenne number being tested for primality. If the Lucas sequence remains divisible by (P), then (P) is a Mersenne prime. This test showcases the intimate connection between Lucas numbers and prime numbers (Tasci & Kilic, 2004) (Figure 7.4).

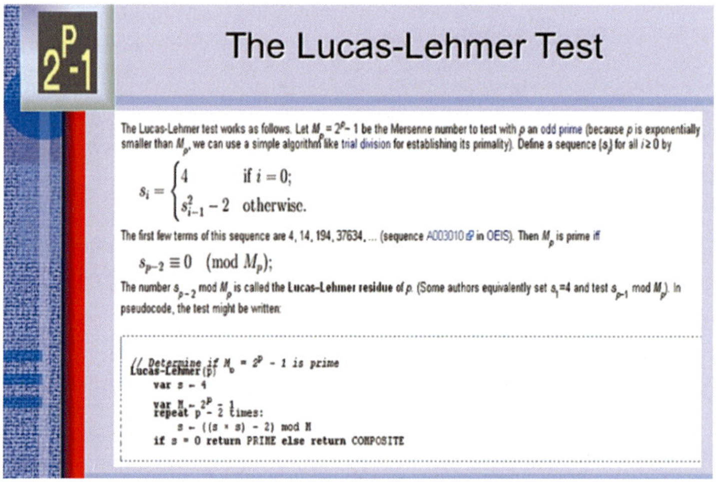

Figure 7.4. The Lucas-Lehmer test.

Source: Ginger Alexander, Creative Commons License.

7.3.2.2. Prime Divisors of Lucas Numbers

Lucas numbers themselves can have prime divisors, adding complexity to their properties. For example, Lucas numbers (L_{11}) and (L_{29}) are both divisible by prime numbers. This can be represented mathematically as:

$$L_{11} = 199 \cdot 521$$

$$L_{29} = 1457 \cdot 491573$$

The presence of prime divisors in Lucas numbers adds an element of number theory and primes to the study of these sequences. It emphasizes the intricate interplay between number theory, sequences, and prime numbers (Deveci & Karaduman, 2012).

7.3.3. In Algebraic Number Theory

Algebraic number theory explores the properties of numbers in algebraic extensions of rational numbers. Lucas numbers find a natural home in this branch of mathematics, contributing to the study of algebraic integers and their properties.

7.3.3.1. Lucas Units in Quadratic Fields

In algebraic number theory, Lucas units play a significant role. These are algebraic integers generated by Lucas numbers in certain quadratic fields. The Lucas-Lehmer test mentioned earlier is intimately related to the properties of Lucas units in quadratic fields. Lucas units exhibit specific properties related to factorization, ideal class groups, and more. This can be represented mathematically as follows:

$$L_n = \frac{á^n - â^n}{á - â}$$

Where $(á)$ and $(â)$ are algebraic integers in the quadratic field.

7.3.3.2. Connection to Pell Equations

Lucas numbers also have connections to the solution of Pell equations. Pell equations are a class of Diophantine equations with solutions that involve Lucas numbers. These equations can be mathematically expressed as:

$$x^2 - Dy^2 = 1$$

Where (D) is a positive integer. The solutions to these equations often involve Lucas numbers and are used in various areas of number theory (Kilic et al., 2010).

7.4. PROPERTIES AND PATTERNS OF HIGHER-ORDER LUCAS SEQUENCES

In this section, we will study the properties and intricate patterns exhibited by higher-order Lucas sequences. These sequences, extensions of the familiar Lucas numbers, reveal interesting arithmetic and geometric properties, exhibit periodicity and cycles, and are governed by recurrence relations tailored to their higher-order nature. Let us explore these aspects in mathematical detail, supported by equations and examples.

7.4.1. Arithmetic and Geometric Properties

Higher-order Lucas sequences introduce arithmetic and geometric properties that extend beyond those of standard Lucas numbers. These properties are defined by the recurrence relation:

$$U_n = PU_{n-1} + QU_{n-2}$$

Here, (P) and (Q) are constants, and (U_0) and (U_1) are initial values.

7.4.1.1. Arithmetic Progressions

Higher-order Lucas sequences form arithmetic progressions, revealing an intriguing relationship between consecutive terms. The common difference in this progression is (P). Mathematically, it can be expressed as:

$$U_{n+1} - U_n = P$$

For example, consider the higher-order Lucas sequence with $(P = 3)$ and $(Q = 1)$ and initial values $(U_0 = 2)$ and $(U_1 = 5)$. The sequence progresses as follows:

2, 5, 17, 56, 185, ...

Each term is 3 greater than the previous term, illustrating the arithmetic progression within the sequence (Falcon & Plaza, 2007).

7.4.1.2. Geometric Growth

Higher-order Lucas sequences also exhibit geometric growth. The ratio between consecutive terms approaches a constant value, which can be expressed as:

$$\lim_{n \to \infty} \frac{U_{n+1}}{U_n} = \frac{P}{Q}$$

For instance, in the same sequence with $(P = 3)$ and $(Q = 1)$, we observe:

$$\frac{U_1}{U_0} = \frac{5}{2} \approx \frac{P}{Q}$$

As (n) increases, the ratio $\left(\frac{U_{n+1}}{U_n}\right)$ approaches $\left(\frac{P}{Q}\right)$, demonstrating the geometric growth of the sequence.

7.4.2. Periodicity and Cycles

Higher-order Lucas sequences can exhibit periodic behavior and cycles, adding depth to their patterns. The presence of cycles is governed by the constants (P) and (Q) and the initial values (Horadam, 1961).

Did You Know?

Lucas sequences, especially in higher orders, can exhibit fascinating periodic and cyclic behavior. This periodicity can be harnessed in various applications, including cryptography and number theory.

7.4.2.1. Periodicity

Periodicity arises when a higher-order Lucas sequence repeats a specific set of terms in a cyclic manner. This phenomenon is determined by the recurrence relation and the initial values. Mathematically, a sequence is periodic if:

$$U_n = U_{n+k}$$

Where (k) is the period of the sequence.

For example, consider the sequence with $(P = 2)$, $(Q = 1)$, and initial values $(U_0 = 1)$ and $(U_1 = 2)$.

The sequence progresses as follows:

$$1, 2, 4, 8, 16, 32, 64, 128, \ldots$$

In this case, the sequence is periodic with a period of 1, as $(U_n = U_{n+1})$ for all (n).

7.4.2.2. Cycles

Cycles occur when a higher-order Lucas sequence returns to a previously encountered set of terms, creating a repeating pattern. The length of the cycle is determined by the constants (P) and (Q) and the initial values. Mathematically, a sequence exhibits a cycle if:

$$U_n = U_{n+k}$$

Where (k) is the length of the cycle.

Consider a sequence with $(P = 3)$, $(Q = 2)$, and initial values $(U_0 = 1)$ and $(U_1 = 3)$. The sequence progresses as follows:

$$1, 3, 9, 27, 81, 243, 729, 2187, 6561, 19683, 59049, \ldots$$

In this case, the sequence exhibits a cycle of length 10, as $(U_{10} = U_0)$, and the pattern repeats (Waddill & Sacks, 1967).

7.4.3. Recurrence Relations for Higher-Order Sequences

The recurrence relations governing higher-order Lucas sequences are tailored to their specific order and characteristics. These recurrence relations provide a systematic way to generate terms in the sequence.

Consider a generalized higher-order Lucas sequence with constants (P) and (Q) and initial values (U_0) and (U_1). The recurrence relation is given by:

$$U_n = PU_{n-1} + QU_{n-2}$$

Where $(n \geq 2)$.

Let's illustrate this with an example:

Suppose we have a higher-order Lucas sequence defined by $(P = 4)$, $(Q = 1)$, and initial values $(U_0 = 2)$ and $(U_1 = 5)$. Using the recurrence relation, we can generate the sequence as follows:

$$U_2 = 4U_1 + 1U_0 = 4 \cdot 5 + 1 \cdot 2 = 22$$

$$U_3 = 4U_2 + 1U_1 = 4 \cdot 22 + 1 \cdot 5 = 93$$

$$U_4 = 4U_3 + 1U_2 = 4 \cdot 93 + 1 \cdot 22 = 394$$

$$U_5 = 4U_4 + 1U_3 = 4 \cdot 394 + 1 \cdot 93 = 1665$$

The recurrence relation systematically generates each term in the sequence based on the previous terms, following the defined constants and initial values (Yayenie, 2011).

7.5. APPLICATIONS OF LUCAS NUMBERS AND HIGHER-ORDER SEQUENCES

In this section, we explore the diverse applications of Lucas numbers and their higher-order counterparts in mathematics and beyond. These applications encompass continued fractions and their approximations, the role of Lucas numbers in pseudoprimes and primality testing, and the fascinating connections between Lucas sequences and geometry in higher dimensions.

7.5.1. Continued Fractions and Approximations

Lucas numbers play a pivotal role in continued fractions, offering precise approximations for various mathematical constants (Catarino, 2014).

Remember

In various mathematical contexts, continued fractions play a significant role. Remember that Lucas numbers and higher-order sequences can be expressed as continued fractions, providing a powerful tool for approximations in different mathematical problems.

7.5.1.1. Continued Fraction Representation

The continued fraction representation of a real number (x) is expressed as:

$$x = a_0 + \cfrac{1}{a_1 + \cfrac{1}{a_2 + \cfrac{1}{a_3 + \cfrac{1}{\dots}}}}$$

Lucas numbers, along with their ratios, can be used to construct continued fractions. Specifically, the continued fraction representation of (f), the golden ratio, involves Lucas numbers:

$$f = 1 + \cfrac{1}{1 + \cfrac{1}{1 + \cfrac{1}{1 + \dots}}} = [1;1,1,1,\dots]$$

The recurrence relation for Lucas numbers $(L_n = L_{n-1} + L_{n-2})$, generates the coefficients $(a_0 = 1)$ and $(a_n = 1)$ for all $(n \geq 1)$, resulting in the continued fraction representation of (f).

7.5.1.2. Approximations with Lucas Numbers

Lucas numbers provide remarkable approximations for (f). By truncating the continued fraction representation at various points, we obtain increasingly accurate approximations:

$$L_0 = 2$$

$$L_1 = 1$$

$$L_2 = 3$$

$$L_3 = 4$$

$$L_4 = 7$$

$$L_5 = 11$$

Using these Lucas numbers as denominators, we can generate the following sequence of approximations for (f):

$$\frac{2}{1}, \frac{3}{2}, \frac{5}{3}, \frac{8}{5}, \frac{13}{8}, \frac{21}{13}, \dots$$

These approximations converge rapidly to the true value of (f), showcasing the effectiveness of Lucas numbers in providing accurate representations of mathematical constants.

7.5.2. Pseudoprimes and Primality Testing

Lucas numbers are intimately connected to pseudoprimes, numbers that exhibit prime-like properties but are not actually prime. These connections have applications in primality testing (Park et al., 2003).

7.5.2.1. Lucas Pseudoprimes

Lucas numbers give rise to a class of pseudoprimes known as Lucas pseudoprimes. A Lucas pseudoprime (U_n) satisfies the Lucas primality test for a given pair of constants (P) and (Q). The test checks whether (U_n) divides evenly into $(2^n - 1)$, where (n) is a positive integer. If the test passes, (U_n) is a Lucas pseudoprime for the specified constants.

For example, consider the Lucas sequence defined by $(P = 1)$ and $(Q = -1)$ with initial values $(U_0 = 0)$ and $(U_1 = 1)$. The sequence progresses as follows:

$$0, 1, -1, 2, -3, 5, -8, 13, -21, 34, \ldots$$

The Lucas primality test examines whether (U_n) divides evenly into $(2^n - 1)$ for various values of (n). If a specific (U_n) satisfies this criterion, it is a Lucas pseudoprime for the given constants (Scott & Marketos, 2014).

7.5.2.2. Applications in Primality Testing

Lucas pseudoprimes have applications in primality testing, especially in cases where the Lucas sequence is chosen strategically. The Lucas-Lehmer test for Mersenne primes, which uses Lucas numbers, is an example of a primality test that relies on the properties of Lucas sequences. This test helps identify Mersenne primes, which are of the form $(2^n - 1)$.

The study of Lucas pseudoprimes highlights the intricate relationship between Lucas numbers and primality testing, contributing to the field of number theory and the identification of prime numbers.

7.5.3. Geometry and Recursion in Higher Dimensions

Lucas sequences extend naturally into higher dimensions, offering intriguing connections to geometry and recursive structures (Atanassov et al., 1985).

7.5.3.1. Higher-Dimensional Lucas Sequences

In higher-dimensional spaces, Lucas sequences can be generalized to include multiple dimensions. Consider a three-dimensional Lucas sequence defined by the following recurrence relation:

$$U_{n+1} = PU_n + QU_{n-1} + RU_{n-2}$$

Here, (P), (Q), and (R) are constants, and (U_0),(U_1), and (U_2) are initial values.

7.5.3.2. Recursive Geometry

Higher-dimensional Lucas sequences are linked to recursive geometric structures. For example, a three-dimensional Lucas sequence can be associated with the recursive generation of three-dimensional objects in geometry. Each term in the sequence influences the position and dimensions of a geometric shape, leading to the creation of complex recursive structures in three dimensions.

7.5.3.3. Applications in Computer Graphics

The recursive geometric properties of higher-dimensional Lucas sequences find applications in computer graphics, where recursive structures are used to model and render three-dimensional objects. The equations governing these sequences allow for the precise placement and scaling of geometric elements, contributing to the realism of computer-generated scenes (Vajda, 2008).

7.6. ADVANCED TOPICS IN LUCAS NUMBERS

In this section, we will study some advanced topics related to Lucas numbers, exploring their connections to Pell equations and quadratic forms, the development of higher-order Lucas polynomials, and their intriguing role in cryptography. These advanced topics highlight the mathematical depth and versatility of Lucas numbers and their applications in various domains.

7.6.1. Connection to Pell Equations and Quadratic Forms

Lucas numbers are intimately linked to Pell equations and quadratic forms, revealing profound mathematical relationships that extend beyond their numerical values.

7.6.1.1. Pell Equations

A Pell equation is a Diophantine equation of the form $\left(x^2 - Dy^2 = 1\right)$, where (D) is a non-square positive integer. Solutions to Pell equations are closely tied to Lucas numbers (Sumeyra et al., 2019).

Let's consider $(D = 2)$ as an example. The Pell equation becomes $\left(x^2 - 2y^2 = 1\right)$, and its solutions are given by (x_n, y_n), where (x_n) and (y_n) are Lucas numbers. Specifically:

$$x_n = L_{2n}$$

$$y_n = L_n$$

For instance, when $(n = 2)$, we have $(x_2 = L_4 = 7)$ and $(y_2 = L_2 = 3)$, which satisfy $\left(7^2 - 2 \cdot 3^2 = 1\right)$, confirming their status as solutions to the Pell equation.

7.6.1.2. Quadratic Forms

Lucas numbers also find relevance in the quadratic forms. Quadratic forms are homogeneous polynomials of degree two in several variables, often used in number theory and algebraic geometry.

One well-known quadratic form is the binary quadratic form $Q(x, y) = x^2 - Dy^2$, where (D) is a non-square positive integer. Lucas numbers can be associated with solutions to this form, providing a means to express them in terms of Lucas numbers (Kilic & Tasci, 2007).

For example, when $(D = 5)$, the binary quadratic form becomes $Q(x, y) = x^2 - 5y^2$. Solutions to this form can be expressed as:

$$x_n = L_{2n}$$

$$y_n = L_{2n-1}$$

Hence, these relationships between Lucas numbers, Pell equations, and quadratic forms underscore the profound mathematical connections that Lucas numbers maintain with fundamental mathematical concepts.

7.6.2. Higher-Order Lucas Polynomials

Higher-order Lucas polynomials are extensions of the familiar Lucas numbers, providing a polynomial representation of Lucas sequences Mestrovic, 2014).

7.6.2.1. Definition

The (n)th higher-order Lucas polynomial, denoted as $L_n(x)$, is defined as follows:

$$L_n(x) = L_{n-1} + xL_{n-2}$$

Here, (n) is a non-negative integer, $(L_{-1} = 2)$, and $(L_0 = 1)$.

7.6.2.2. Properties

Higher-order Lucas polynomials exhibit several intriguing properties. They can be expressed as a polynomial in terms of (x), allowing for efficient computation and manipulation. These polynomials are closely related to the Lucas numbers, and their coefficients are derived from Lucas sequences.

For instance, the first few higher-order Lucas polynomials are:

$$L_0(x) = 1$$

$$L_1(x) = x$$

$$L_2(x) = x^2 + 1$$

$$L_3(x) = x^3 + 2x$$

$$L_4(x) = x^4 + 3x^2 + 1$$

These polynomials extend the properties and patterns observed in Lucas numbers, providing a polynomial representation of Lucas sequences that finds applications in algebra and number theory.

7.6.3. Lucas Numbers in Cryptography

Lucas numbers, with their unique properties, have found applications in cryptographic algorithms, contributing to the security and efficiency of encryption and decryption processes (Hou, 1987).

7.6.3.1. Lucas Pseudoprimes in Cryptography

Lucas pseudoprimes, as mentioned earlier, are numbers that pass Lucas primality tests. In cryptography, these pseudoprimes are utilized in probabilistic primality tests, enhancing the security of encryption algorithms. The Lucas-Lehmer primality test, which leverages Lucas pseudoprimes, plays a role in identifying large prime numbers used in cryptographic protocols.

7.6.3.2. Pseudorandom Number Generation

Lucas sequences can be employed in pseudorandom number generation, a fundamental component of cryptographic algorithms. By selecting appropriate initial values and constants, Lucas sequences can produce pseudorandom sequences that exhibit desirable statistical properties, making them suitable for encryption and decryption processes (Kusljevic, 2004).

7.6.3.3. Cryptographic Protocols

Lucas numbers also find applications in cryptographic protocols, including key exchange and digital signatures. Their role in these protocols is to provide a secure and efficient foundation for data encryption and authentication (Helland et al., 1992).

Activity 7.1

Imagine you are designing a pseudorandom number generator for a computer game. How could you utilize Lucas numbers or higher-order Lucas sequences to enhance the randomness and unpredictability of the generated numbers while maintaining computational efficiency?

7.7. SUMMARY

This chapter discussed Lucas numbers and their higher-order counterparts. It begins with a comprehensive introduction, explaining the definition, properties, and their close relationship with the well-known Fibonacci numbers. The chapter discusses generalized Lucas sequences, extending into higher orders, offering recursive definitions and initial conditions.

Furthermore, it unravels the widespread occurrence of Lucas numbers in both the natural world and mathematics, illustrating their connection to prime numbers and their role in algebraic number theory. The chapter delves into the properties and patterns inherent in higher-order Lucas sequences, discussing their arithmetic, geometric properties, periodicity, and recurrence relations.

Applications of Lucas numbers and higher-order sequences are unveiled, including their utilization in continued fractions for approximations, pseudoprime identification, and even geometry in higher dimensions. The chapter also delves into advanced topics, such as their link to Pell equations, higher-order Lucas polynomials, and their relevance in cryptography.

REFERENCES

1. Abbad, S., Belbachir, H., & Benzaghou, B., (2019). Companion sequences associated to the $ r $-Fibonacci sequence: Algebraic and combinatorial properties. *Turkish Journal of Mathematics, 43*(3), 1095–1114.
2. Atanassov, K. T., Atanassova, L., & Sasselov, D., (1985). A new perspective to the generalization of the Fibonacci sequence. *The Fibonacci Quarterly, 23*(1), 21–28.
3. Baillie, R., & Wagstaff, S. S., (1980). Lucas pseudoprimes. *Mathematics of Computation, 35*(152), 1391–1417.
4. Ballabio, C., Panagos, P., & Monatanarella, L., (2016). Mapping topsoil physical properties at European scale using the LUCAS database. *Geoderma, 261*, 110–123.
5. Barnsley, M. F., Hutchinson, J. E., & Stenflo, Ö., (2008). V-variable fractals: Fractals with partial self-similarity. *Advances in Mathematics, 218*(6), 2051–2088.
6. Battaloglu, R., & Simsek, Y., (2021). On new formulas of Fibonacci and Lucas numbers involving golden ratio associated with atomic structure in chemistry. *Symmetry, 13*(08), 1334.
7. Bosma, W., & Van, D. H. M. P., (1989). Faster primality testing. In: *Workshop on the Theory and Application of Cryptographic Techniques* (Vol. 1, No. 1, pp. 652–656). Berlin, Heidelberg: Springer Berlin Heidelberg.
8. Boussayoud, A., (2017). On some identities and generating functions for Pell-Lucas numbers. *Online. J. Anal. Comb., 12*(1), 1–10.
9. Bravo, J. J., Herrera, J. L., & Luca, F., (2021). On a generalization of the Pell sequence. *Mathematica Bohemica, 146*(2), 199–213.
10. Bruce, J. W., (1993). A really trivial proof of the Lucas-Lehmer primality test. *The American Mathematical Monthly, 100*(4), 370–371.
11. Castro, M., Gallardo, J., & Parés, C., (2006). High order finite volume schemes based on reconstruction of states for solving hyperbolic systems with nonconservative products. Applications to shallow-water systems. *Mathematics of Computation, 75*(255), 1103–1134.
12. Catarino, P., & Vasco, P., (2020). The generalized order-m (k-Pell) numbers. *Annals of the Alexandru Ioan Cuza University-Mathematics, 66*(1).

13. Catarino, P., (2014). On some identities for k-Fibonacci sequence. *Int. J. Contemp. Math. Sci, 9*(1), 37–42.
14. Çelik, S., Durukan, İ., & Özkan, E., (2021). New recurrences on Pell numbers, Pell-Lucas numbers, Jacobsthal numbers, and Jacobsthal-Lucas numbers. *Chaos, Solitons & Fractals, 150*, 111173.
15. Chen, H., & Wang, Y., (2011). A family of higher-order convergent iterative methods for computing the Moore–Penrose Inverse. *Applied Mathematics and Computation, 218*(8), 4012–4016.
16. Cohen, H., & Lenstra, A. K., (1987). Implementation of a new primality test. *Mathematics of Computation, 48*(177), 103–121.
17. Coquand, T., & Huet, G., (1985). Constructions: A higher order proof system for mechanizing mathematics. In: *European Conference on Computer Algebra* (pp. 151–184). Berlin, Heidelberg: Springer Berlin Heidelberg.
18. Deveci, Ö., & Karaduman, E., (2012). The generalized order-k Lucas sequences in finite groups. *Journal of Applied Mathematics, 2012*, 1–15.
19. Dresden, G. P., & Du, Z., (2014). A simplified Binet formula for k-generalized Fibonacci numbers. *J. Integer Seq., 17*(4), 4–14.
20. Ediriweera, S., Jayarathna, P., Lucas, R., & Samarasinghe, R., (2019). Fractal analysis of creeping discharges propagating over solid/liquid interfaces-influence of oil level. In: *2019 9th International Conference on Power and Energy Systems (ICPES)* (Vol. 1, No. 1, pp. 1–6). IEEE.
21. Edson, M., & Yayenie, O., (2009). *A New Generalization of Fibonacci Sequence & Extended Binet's Formula, 1*(1), 1–20.
22. Erduvan, F., Keskin, R., & Şiar, Z., (2021). Repdigits base b as products of two Pell numbers or Pell–Lucas numbers. *Boletín de la Sociedad Matemática Mexicana, 27*, 1–17.
23. Falcon, S., & Plaza, Á., (2007). The k-Fibonacci sequence and the Pascal 2-triangle. *Chaos, Solitons & Fractals, 33*(1), 38–49.
24. Falcon, S., (2011). On the k-Lucas numbers. *International Journal of Contemporary Mathematical Sciences, 6*(21), 1039–1050.
25. Fuentes, F., Keith, B., Demkowicz, L., & Nagaraj, S., (2015). Orientation embedded high order shape functions for the exact sequence elements of all shapes. *Computers & Mathematics with Applications, 70*(4), 353–458.

26. Gökbas, H., & Köse, H., (2017). Some sum formulas for products of Pell and Pell-Lucas numbers. *Int. J. Adv. Appl. Math. and Mech, 4*(4), 1–4.

27. Gulec, H. H., Taskara, N., & Uslu, K., (2013). A new approach to generalized Fibonacci and Lucas numbers with binomial coefficients. *Applied Mathematics and Computation, 220*, 482–486.

28. Hata, M., (1986). Fractals in mathematics. In: *Studies in Mathematics and its Applications* (Vol. 18, pp. 259–278). Elsevier.

29. Helland, K., Berntsen, H. E., Borgen, O. S., & Martens, H., (1992). Recursive algorithm for partial least squares regression. *Chemometrics and Intelligent Laboratory Systems, 14*(1–3), 129–137.

30. Horadam, A. F., (1961). A generalized Fibonacci sequence. *The American Mathematical Monthly, 68*(5), 455–459.

31. Hou, H., (1987). A fast recursive algorithm for computing the discrete cosine transform. *IEEE Transactions on Acoustics, Speech, and Signal Processing, 35*(10), 1455–1461.

32. Husain, A., Nanda, M. N., Chowdary, M. S., & Sajid, M., (2022). Fractals: An eclectic survey, part II. *Fractal and Fractional, 6*(7), 379.

33. Hutchinson, J. E., (1981). Fractals and self-similarity. *Indiana University Mathematics Journal, 30*(5), 713–747.

34. Ibrahim, M., (2023). On the eight levels theorem and applications towards Lucas-Lehmer primality test for Mersenne primes, I. *Arab Journal of Basic and Applied Sciences, 30*(1), 267–284.

35. Jelinek, H. F., Jones, C. L., Warfel, M. D., Lucas, C., Depardieu, C., & Aurel, G., (2006). Understanding fractal analysis? The case of fractal linguistics. *Complexus, 3*(1–3), 66–73.

36. Kiliç, E., & Taşci, D., (2006). On the generalized order-k Fibonacci and Lucas numbers. *The Rocky Mountain Journal of Mathematics*, 1915–1926.

37. Kilic, E., & Taşci, D., (2007). On the permanents of some tridiagonal matrices with applications to the Fibonacci and Lucas numbers. *The Rocky Mountain Journal of Mathematics, 1*(1), 1953–1969.

38. Kilic, E., (2007). The generalized order-k Fibonacci–Pell sequence by matrix methods. *Journal of Computational and Applied Mathematics, 209*(2), 133–145.

39. Kilic, E., (2008). The Binet formula, sums, and representations of generalized Fibonacci p-numbers. *European Journal of Combinatorics, 29*(3), 701–711.

40. Kılıç, E., (2009). The generalized Pell (p, i)-numbers and their Binet formulas, combinatorial representations, sums. *Chaos, Solitons & Fractals, 40*(4), 2047–2063.
41. Kilic, E., Tasci, D., & Haukkanen, P., (2010). On the generalized Lucas sequences by Hessenberg matrices. *ARS Combin., 95*, 383–395.
42. Kohlenbach, U., (2005). Higher order reverse mathematics. In: *Reverse Mathematics 2001* (pp. 281–295). AK Peters/CRC Press.
43. Koken, F., & Bozkurt, D., (2008). On the Jacobsthal-Lucas numbers by matrix methods. *Int. J. Contemp. Math. Sciences, 3*(33), 1629–1633.
44. Koshy, T., (2014). *Pell and Pell-Lucas Numbers with Applications* (Vol. 431), 1–20.
45. Koshy, T., (2019). *Fibonacci and Lucas Numbers with Applications* (Vol. 1, 2, No. 1, pp. 1–20). John Wiley & Sons.
46. Kusljevic, M. D., (2004). A simple recursive algorithm for frequency estimation. *IEEE Transactions on Instrumentation and Measurement, 53*(2), 335–340.
47. Lathrop, J. I., Lutz, J. H., & Summers, S. M., (2009). Strict self-assembly of discrete Sierpinski triangles. *Theoretical Computer Science, 410*(4, 5), 384–405.
48. Lehmer, D. H., (1935). On Lucas's test for the primality of Mersenne's numbers. *Journal of the London Mathematical Society, 1*(3), 162–165.
49. Lengyel, T., (1995). The order of the Fibonacci and Lucas numbers. *Fibonacci Quart, 33*(3), 234–239.
50. Leyendekkers, J. V., & Shannon, A. G., (2016). Some golden ratio generalized Fibonacci and Lucas sequences. *Notes on Number Theory and Discrete Mathematics, 22*(1), 33–41.
51. Li, C., Shen, Y., Ge, H., Su, S., & Yang, Z., (2016). Analysis of spontaneous imbibition in fractal tree-like network system. *Fractals, 24*(3), 1650035.
52. Li, C., Zhang, X., Li, N., Wang, Y., Yang, J., Gu, G., & Wang, Y., (2017). Construction of Sierpinski triangles up to the fifth order. *Journal of the American Chemical Society, 139*(39), 13749–13753.
53. Menken, H., & Dişkaya, O., (2019). On the quadra Fibona-Pell and Hexa Fibona-Pell-Jacobsthal sequences. *Mathematical Sciences and Applications E-Notes, 7*(2), 149–160.
54. Meštrović, R., (2014). *Lucas' Theorem: Its Generalizations, Extensions and Applications* (Vol. 1, pp. 1–43)

55. Noriyuki, S. U. W. A., (2020). Erratum to geometric aspects of Lucas sequences, I. *Tokyo Journal of Mathematics, 43*(2), 543–544.
56. Panwar, Y. K., Singh, B., & Gupta, V. K., (2013). Identities of common factors of generalized Fibonacci, Jacobsthal and Jacobsthal-Lucas numbers. *Applied Mathematics and Physics, 1*(4), 126–128.
57. Park, A. E., Fernandez, J. J., Schmedders, K., & Cohen, M. S., (2003). The Fibonacci sequence: Relationship to the human hand. *The Journal of Hand Surgery, 28*(1), 157–160.
58. Peitgen, H. O., Jürgens, H., Saupe, D., Peitgen, H. O., Jürgens, H., & Saupe, D., (1992). Classical fractals and self-similarity. *Chaos and Fractals: New Frontiers of Science,* 63–134.
59. Pinto, C. M., Lopes, A. M., & Machado, J. T., (2014). Double power laws, fractals, and self-similarity. *Applied Mathematical Modelling, 38*(15, 16), 4019–4026.
60. Ramírez, J. L., (2015). Some combinatorial properties of the k-Fibonacci and the k-Lucas quaternions. *Analele Ştiinţifice ale Universităţii "Ovidius" Constanţa. Seria Matematică, 23*(2), 201–212.
61. Razpet, M., (2002). The Lucas property of a number array. *Discrete Mathematics, 248*(1–3), 157–168.
62. Roesler, U., & Rüschendorf, L., (2001). The contraction method for recursive algorithms. *Algorithmica, 29*, 3–33.
63. Schöberl, J., & Zaglmayr, S., (2005). High order Nédélec elements with local complete sequence properties. *Compel-the International Journal for Computation and Mathematics in Electrical and Electronic Engineering, 24*(2), 374–384.
64. Scott, T. C., & Marketos, P., (2014). On the origin of the Fibonacci sequence. *MacTutor History of Mathematics, 23*, 1–20.
65. Shang, J., Wang, Y., Chen, M., Dai, J., Zhou, X., Kuttner, J., & Wu, K., (2015). Assembling molecular Sierpiński triangle fractals. *Nature Chemistry, 7*(5), 389–393.
66. Siar, Z., & Keskin, R., (2014). The square terms in generalized Lucas sequences. *Mathematika, 60*(1), 85–100.
67. Singh, B., Sikhwal, O., & Bhatnagar, S., (2010). Fibonacci-Like sequence and its properties. *Int. J. Contemp. Math. Sciences, 5*(18), 859–868.
68. Singh, M., Gupta, Y., & Sikhwal, O., (2014). Identities of generalized Fibonacci-like sequence. *Turkish Journal of Analysis and Number Theory, 2*(5), 170–175.

69. Soykan, Y., (2021). Binomial transform of the generalized third order Pell sequence. *Communications in Mathematics and Applications, 12*(1), 71.
70. Staples, M. E., & Truxaw, M. P., (2012). An initial framework for the language of higher-order thinking mathematics practices. *Mathematics Education Research Journal, 24*, 257–281.
71. Strichartz, R. S., (1999). Analysis on fractals. *Notices AMS, 46*(10), 1199–1208.
72. Sümeyra, U. Ç. A. R., Nihal, T. A. Ş., & Özgür, N. Y., (2019). A new application to coding theory via Fibonacci and Lucas numbers. *Mathematical Sciences and Applications E-Notes, 7*(1), 62–70.
73. Szynal-Liana, A., (2018). The Horadam hybrid numbers. *Discussiones Mathematicae: General Algebra & Applications, 3*(1), 1–15.
74. Tasci, D., & Kilic, E., (2004). On the order-k generalized Lucas numbers. *Applied Mathematics and Computation, 155*(3), 637–641.
75. Thompson, T., (2008). Mathematics teachers' interpretation of higher-order thinking in Bloom's taxonomy. *International Electronic Journal of Mathematics Education, 3*(2), 96–109.
76. Vajda, S., (2008). *Fibonacci and Lucas Numbers, and the Golden Section: Theory and Applications* (Vol. 1, No. 1, pp. 1–20). Courier Corporation.
77. Volk, M., Cotič, M., Zajc, M., & Starcic, A. I., (2017). Tablet-based cross-curricular math's vs. traditional math's classroom practice for higher-order learning outcomes. *Computers & Education, 114*, 1–23.
78. Waddill, M. E., & Sacks, L., (1967). Another generalized Fibonacci sequence. *The Fibonacci Quarterly, 5*(3), 209–222.
79. Yağmur, T., (2019). New approach to Pell and Pell-Lucas sequences. *Kyungpook Mathematical Journal, 59*(1), 23–34.
80. Yayenie, O., (2011). A note on generalized Fibonacci sequences. *Applied Mathematics and Computation, 217*(12), 5603–5611.
81. Yayenie, O., (2012). New identities for generalized Fibonacci sequences and new generalization of lucas sequences. *Southeast Asian Bulletin of Mathematics, 36*(5), 1–20.
82. Yilmaz, F., & Bozkurt, D., (2009). The generalized order-k Jacobsthal numbers. *Int. J. Contemp. Math. Sciences, 4*(34), 1685–1694.
83. Zhang, X., & Han, Y., (2012). Existence and uniqueness of positive solutions for higher order nonlocal fractional differential equations. *Applied Mathematics Letters, 25*(3), 555–560.

CHAPTER 8

Beyond Basics: Advanced Number Patterns

Contents

8.1. INTRODUCTION

In this chapter, we lay the foundation for the advanced concepts that will follow, setting the stage for a profound mathematical exploration (Athanasiadis et al., 2021). These advanced patterns are not only intellectually stimulating but also possess profound importance in various branches of science, engineering, and cryptography, where mathematical concepts are utilized in real-world applications. Our pursuit of advanced number patterns is driven by the intrinsic features of mathematical structures that transcend the ordinary. As we go deeper into this domain, we shall encounter mathematical objects and relationships that challenge our understanding and expand the boundaries of our mathematical knowledge. To understand the importance of exploring these deeper mathematical structures, one need only consider their far-reaching applications. From cryptography to quantum computing, these advanced number patterns are used in modern technology and scientific inquiry. They are the hidden forces that drive the secure transmission of information, the optimization of algorithms, and the decryption of secrets (Le et al., 2022).

For instance, the concept of Lucas numbers which are intimately related to the famous Fibonacci sequence. These numbers are defined recursively as $L(n) = L(n-1) + L(n-2)$, with initial values of $L(0) = 2$ and $L(1) = 1$. However, they possess deep connections to number patterns, number theory, Diophantine equations, and even the study of elliptic curves. As we explore Lucas numbers, we will uncover their remarkable properties and their role in various mathematical landscapes (Lin et al., 2021). Farey sequences, comprising the rational numbers between 0 and 1 with reduced fractions, reveal a hidden order in the chaotic world of real numbers. Farey sequences find applications in number theory, continued fractions, and even the study of chaotic dynamical systems. In the coming sections of this chapter, we will delve into Möbius transformations, partition theory, chaos theory, transcendental numbers, prime number distributions, quantum computing, and more. Each topic will be a gateway to a world of mathematical beauty and practical significance (Murphy et al., 2022).

8.2. LUCAS NUMBERS AND THEIR PROPERTIES

In our exploration of advanced number patterns, Lucas numbers stand as an intriguing mathematical entity. These numbers, while lesser known than their Fibonacci counterparts, exhibit unique properties and connections that make them a compelling subject of study (Son, 2014).

8.2.1. Properties and Connections to Fibonacci Numbers

Lucas numbers share several fascinating properties with the more well-known Fibonacci sequence. One remarkable connection between the two is expressed by the relationship:

$$L(n) = F(n-1) + F(n+1)$$

Here, $F(n)$ represents the nth term of the Fibonacci sequence. This connection underscores the inherent symmetry between Lucas and Fibonacci numbers, emphasizing their parallel nature (Benavoli et al., 2009).

Additionally, the ratio of consecutive Lucas numbers, $L(n+1)/L(n)$, approaches the golden ratio $(f \approx 1.6180339887)$ as (n) increases. This convergence to the golden ratio is similar to the behavior of the ratio of consecutive Fibonacci numbers. It is a testament to the pervasive influence of the golden ratio in number patterns. Lucas numbers also play a crucial role in the study of Diophantine equations, which are equations with integer solutions. They appear in the context of elliptic curves, further highlighting their significance in number theory (Krcadinac, 2006). In conclusion, Lucas numbers, defined by their recursive nature and specific initial values, offer a captivating journey into the world of advanced number patterns. Their intriguing properties and connections to the Fibonacci sequence demonstrate the profound interconnectedness of mathematical structures. As we delve deeper into the realm of advanced number patterns, Lucas numbers serve as a testament to the rich tapestry of mathematics, where seemingly distinct threads are woven together into a unified fabric of numerical elegance (Figure 8.1).

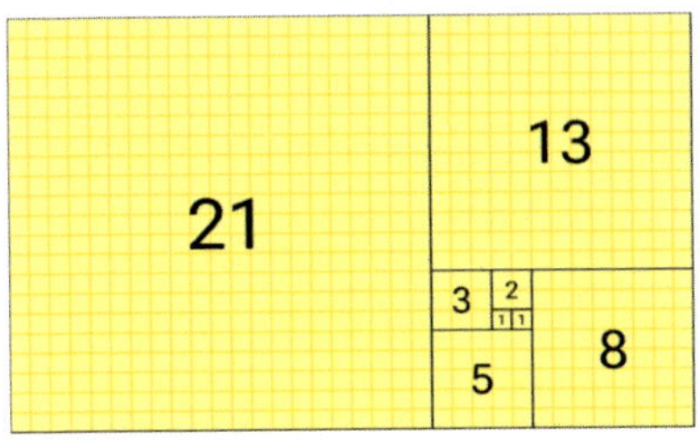

Figure 8.1. Illustration of Lucas number's connections to Fibonacci.

Source: Romain, Creative Commons License.

8.3. FAREY SEQUENCES AND NUMBER THEORY

In our exploration of advanced number patterns, Farey sequence pentatopes emerge as an interesting topic with profound connections to number theory and Diophantine equations. This section will discuss the understanding of Farey sequences, their applications in number theory and Diophantine equations, and their mathematical properties and significance (Schielack, 1987).

8.3.1. Understanding Farey Sequences

Farey sequences are sets of rational numbers that lie within a specific interval [0, 1], arranged in ascending order and reduced to their simplest form. These sequences are named after the British geologist John Farey, who studied them in the early 19th century.

To construct a Farey sequence of order n, one starts with the interval [0, 1] and divides it into n equal parts. Then, all the fractions with denominators less than or equal to n are considered, ensuring that each fraction is in its reduced form. The resulting sequence, denoted as F(n), contains all the unique fractions within [0, 1] (van Gend, 2014).

For instance, the Farey sequence of order 5, denoted as F (5), includes the fractions 0/1, 1/5, 1/4, 1/3, 2/5, 1/2, 3/5, 2/3, 3/4, 4/5, and 1/1. These fractions represent a systematic exploration of rational numbers in the interval [0, 1] (Figure 8.2).

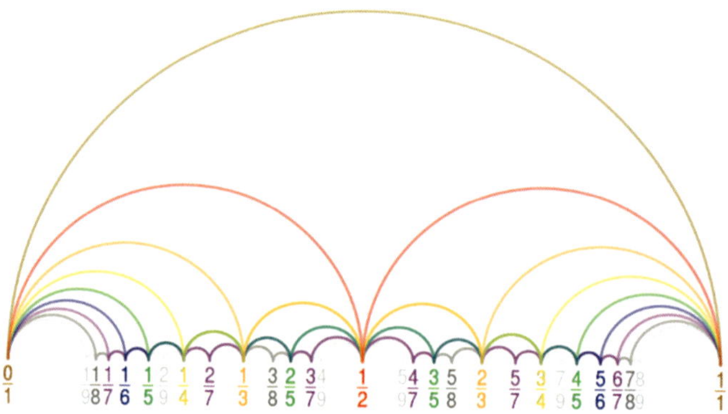

Figure 8.2. Schematic of Farey sequence.

Source: Wikimedia, Creative Commons License.

8.3.2. Applications in Number Theory and Diophantine Equations

Farey sequences find valuable applications in number theory and the study of Diophantine equations, which involve finding integer solutions to polynomial equations. They provide a framework for understanding the distribution of rational numbers, particularly in relation to irreducible fractions (Leroy et al., 2016).

One notable application is in approximating irrational numbers. Given an irrational number α, Farey sequences help find the best rational approximations by considering fractions within the sequences that come closest to α. This is especially significant in continued fraction expansions, where Farey sequences aid in approximating the values of quadratic irrationals.

Furthermore, Farey sequences have applications in the study of Diophantine equations, such as Pell's equation. By systematically exploring fractions within Farey sequences, mathematicians can identify solutions to Diophantine equations and understand the properties of their solutions Granville, 1997).

8.3.3. Mathematical Properties and Significance

Farey sequences possess intriguing mathematical properties that make them a subject of interest in number pattern and number theory. One such property is the Farey sum, which involves combining two fractions from a Farey sequence to create a new fraction.

The sum of two adjacent fractions in a Farey sequence results in another fraction that lies between them. This property has implications for continued fraction expansions and the study of irrational numbers. Another significant property is related to the density of fractions within Farey sequences. As the order of the sequence increases, the density of fractions approaches a constant known as the Farey constant.

This behavior contributes to our understanding of the distribution of rational numbers and their relationships with irrational numbers (de Weger, 1997).In conclusion, Farey sequences offer a profound exploration of rational numbers within the interval [0, 1]. Their applications in number pattern, number theory, and Diophantine equations, coupled with their mathematical properties, underscore their significance in advanced number patterns. Farey sequences serve as a valuable tool for approximating irrationals, studying

Diophantine equations, and highlighting the intricate connections between rational and irrational numbers in the realm of mathematics.

8.4. MÖBIUS TRANSFORMATIONS AND COMPLEX ANALYSIS

In our exploration of advanced number patterns, we encounter Möbius transformations, a subject linked with complex analysis. This section will provide an introduction to Möbius transformations, explaining the pivotal role of complex numbers in this context and highlighting their applications in conformal mapping and geometry (Hoggatt, 1967).Möbius transformations, named after the German mathematician August Ferdinand Möbius, are a class of functions in complex analysis that play a crucial role in mapping one region of the complex plane to another. They are defined as functions of the form:

$$f(z) = \frac{az + b}{cz + d},$$

Where $(a), (b), (c),$ and (d) are complex numbers satisfying $ad - bc \neq 0$. The domain and range of Möbius transformations are the extended complex plane, including the point at infinity (Benavoli et al., 2009). One of the remarkable features of Möbius transformations is their ability to preserve angles between curves. This property, known as conformality, makes them invaluable in various mathematical and geometric contexts (Figure 8.3).

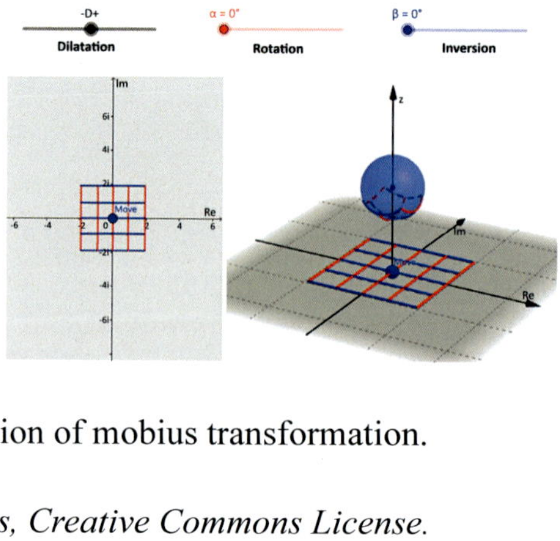

Figure 8.3. Illustration of mobius transformation.

Source: Juan Carlos, Creative Commons License.

8.4.1. The Role of Complex Numbers

Complex numbers are fundamental to the theory of Möbius transformations. The complex plane provides a geometric setting where Möbius transformations can be visualized and understood. Complex numbers are used to represent points in the plane, making Möbius transformations a powerful tool for manipulating and transforming complex functions (Krcadinac, 2006).

Möbius transformations leverage the algebraic and geometric properties of complex numbers to achieve their mapping capabilities. The denominator in the Möbius transformation formula is carefully designed to ensure that certain important properties are preserved during the transformation process.

8.4.2. Applications in Conformal Mapping and Geometry

One of the central applications of Möbius transformations is in conformal mapping, a branch of complex analysis concerned with preserving angles and local properties of curves. Möbius transformations are particularly adept at this task due to their angle-preserving nature.

Remember

Understanding complex numbers and Möbius transformations is crucial in the study of conformal mapping and geometry. Keep in mind that these mathematical concepts are powerful tools for solving intricate geometric problems.

In geometry, Möbius transformations are used to map circles and lines to other circles and lines. This property is essential for solving problems involving circles and lines in complex geometry.

Additionally, Möbius transformations find applications in various areas, including fluid dynamics, electrical engineering, and computer graphics. They are employed to model and analyze the behavior of physical systems, design electronic circuits, and create computer-generated images with intricate shapes (Schielack, 1987).

In summary, Möbius transformations are a captivating topic within advanced number patterns, intimately connected with complex analysis. These transformations, defined through complex numbers, have the remarkable ability to preserve angles and local properties of curves. Their applications in conformal mapping, geometry, and various scientific and engineering disciplines underscore their significance in the realm of mathematics and beyond.

8.5. PARTITION THEORY AND INTEGER COMPOSITIONS

In advanced number patterns, partition theory and the concept of integer compositions provide a direction through which we can explore the ways integers can be broken down into smaller components. This section discusses partition theory, highlighting the counting of integer compositions and showcasing their wide-ranging applications in combinatorics and number theory (Granville, 1997).

8.5.1. Exploring Partition Theory

Partition theory is a branch of number theory that deals with the ways in which a positive integer can be expressed as the sum of smaller positive integers. These expressions are called partitions. For instance, the number 5 can be partitioned in various ways, such as 5, 4+1, 3+2, 3+1+1, 2+2+1, and 2+1+1+1. The study of partition theory involves understanding the patterns and properties of these partitions.A fundamental concept in partition theory is the partition function, denoted by $p(n)$, which represents the number of distinct partitions of the integer (n). Euler's pentagonal number theorem is a notable result in this field, providing a recursive formula for calculating $p(n)$.

8.5.2. Counting Integer Compositions

Integer compositions are a related concept, albeit with a distinct focus. In compositions, we break down a positive integer into smaller positive integers while considering the order of the components. For example, the integer 5 can be composed as (1, 1, 1, 1, 1), (2, 1, 1, 1), (1, 2, 1, 1), and so on. Compositions offer a different perspective on integer breakdown, emphasizing the sequential arrangement of the components (de Weger, 1997).

To count the number of compositions of an integer (n) into (k) parts, we use combinatorial techniques, often involving generating functions and combinatorial identities. The study of integer compositions provides valuable insights into combinatorics and combinatorial number theory.

8.5.3. Applications in Combinatorics and Number Theory

The significance of partition theory and integer compositions extends beyond their intrinsic mathematical feature. They find applications in various branches of mathematics, particularly in combinatorics and number theory.

In combinatorics, partition theory is used to solve problems related to counting and arranging objects with specific constraints. It provides a systematic approach to tackling combinatorial problems involving partitions of sets, permutations, and more.

In number theory, partition theory has connections to modular forms and the theory of elliptic functions. The partition function $p(n)$ has fascinating connections to various mathematical areas, including complex analysis and algebraic number theory.

Integer compositions are integral to combinatorial number theory, where they are employed to investigate the properties of integers and their representations. They are used to solve problems related to partitions of integers into summands with specific properties (Hoggatt, 1967) (Figure 8.4).

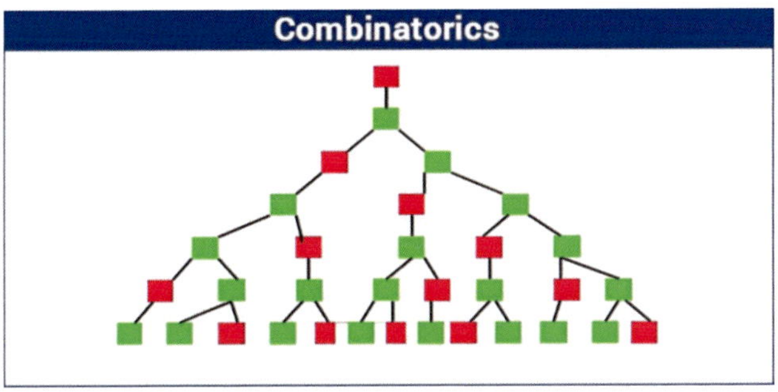

Figure 8.4. Image of combinatorics.

Source: Byju's, Creative Commons License.

In conclusion, partition theory and integer compositions are captivating topics within advanced number patterns. These concepts allow us to explore the diverse ways integers can be expressed as sums or compositions of smaller integers. Their applications in combinatorics and number theory underscore their importance in the realm of mathematics and their relevance in solving a wide range of mathematical problems.

8.6. CHAOS THEORY AND FRACTALS IN NUMBER PATTERNS

The connection between chaos theory and fractals in number patterns provides a unique perspective on seemingly random and complex sequences. This section explores the emergence of chaos in number sequences, the concept of fractals and self-similarity, and the profound connection between chaos theory and number patterns (Konvalina, 2000).

8.6.1. The Emergence of Chaos in Number Sequences

Chaos theory, a branch of mathematics that deals with systems sensitive to initial conditions, has found unexpected applications in the study of number patterns. In certain instances, seemingly orderly number sequences can exhibit chaotic behavior, defying predictability. This chaos arises from the intricate relationships between numbers within the sequence.

One classic example of chaos in number sequences is the logistic map, which describes the population dynamics of a species with limited resources. As the rate of reproduction varies, the population can exhibit erratic, unpredictable behavior, even though the underlying equations are deterministic. This chaotic behavior is of great interest in understanding complex systems and their inherent unpredictability (Annamalai, 2022).

8.6.2. Fractals and Self-Similarity

These complex geometric structures exhibit self-similarity, meaning that as we zoom in on a small portion of the fractal, it resembles the entire structure. The idea of self-similarity is integral to fractals, making them a powerful tool for modeling and visualizing chaotic systems.

One well-known fractal is the Mandelbrot set, which arises from iterating a simple mathematical formula. The resulting image is a breathtaking display of intricate patterns and self-similarity on different scales. The Mandelbrot set's boundaries reveal an infinite complexity that underscores the unpredictable nature of chaotic systems (Bateman & Horn, 1962).

8.6.3. Chaos Theory and Number Patterns

The relationship between chaos theory and number patterns is not coincidental. Chaotic systems often emerge from simple mathematical operations, such as iterating a formula. This inherent simplicity gives rise to complex, irregular behaviors that are reflected in the resulting number sequences.

For example, consider the logistic map once more. It involves iterating a straightforward formula to model population growth. Yet, under certain conditions, this map can produce chaotic sequences of population sizes. The seemingly unpredictable nature of these sequences highlights the subtle ways in which chaos can manifest within number patterns.

The study of chaos theory and fractals in number patterns has applications in various fields, including physics, biology, and finance. It allows researchers to gain insights into the behavior of complex systems that are otherwise challenging to predict. Additionally, it challenges our conventional understanding of order and randomness, demonstrating that chaos can emerge from simplicity (Soundararajan, 2007).

In conclusion, chaos theory and fractals offer a captivating perspective on number patterns. The emergence of chaos in number sequences, coupled with the self-similar and intricate nature of fractals, creates an interesting mathematical phenomenon. This connection between chaos theory and number patterns opens doors to new insights and applications across multiple domains, making it a compelling area of study in advanced number theory.

8.7. TRANSCENDENTAL NUMBERS AND CONTINUED FRACTIONS

In advanced number patterns, transcendental numbers and continued fractions offer profound insights into the unique and fascinating properties of certain real numbers. In this section, we will discuss the definition of transcendental numbers, explore the utility of continued fractions, and unravel the uniqueness and properties associated with these representations (Karatsuba, 1990).

8.7.1. Defining Transcendental Numbers

Transcendental numbers are a class of real numbers that are neither algebraic nor rational. In other words, they cannot be expressed as solutions to any non-zero polynomial equation with integer coefficients. While transcendental numbers may seem elusive, they have significant mathematical importance. Notable examples of transcendental numbers include π (pi) and e (Euler's number).

Did You Know?

Transcendental numbers are a rare and unique class of numbers. Did you know that π (pi) and e (Euler's number) are examples of transcendental numbers? They cannot be expressed as roots of any non-zero polynomial equation with integer coefficients.

Transcendental numbers are characterized by their non-repeating, non-terminating decimal expansions. This property distinguishes them from algebraic numbers, which can be expressed as solutions to polynomial equations. The transcendence of certain numbers, like π and e, has been proven using advanced mathematical techniques, and they continue to be objects of fascination and study in number theory.

8.7.2. The Use of Continued Fractions

Continued fractions provide a unique way to represent real numbers as an infinite sequence of fractions. These sequences capture the essence of real numbers with remarkable precision. The utility of continued fractions lies in their ability to approximate real numbers efficiently and express them in a compact form (Shparlinski, 2008).

A continued fraction is typically represented as $[a_0; a_1, a_2, a_3,.]$, where a_0 is an integer and $a_1, a_2, a_3,.$ are positive integers. The sequence extends indefinitely, representing a real number accurately. Continued fractions can be finite or infinite, depending on the nature of the real number being represented.

8.7.3. Properties of Continued Fraction Representations

One of the remarkable aspects of continued fractions is their uniqueness. Every real number has a unique continued fraction representation. This uniqueness is a testament to the power of continued fractions in capturing the essential properties of real numbers.

The process of finding the continued fraction representation of a real number involves recursively calculating the integer and fractional parts of the number.

Each term in the sequence provides increasingly accurate approximations of the real number. This makes continued fractions an invaluable tool in approximating and understanding the mathematical properties of transcendental numbers (Carsault et al., 2021).

For instance, consider the golden ratio, often denoted by φ (phi). Its continued fraction representation is [1; 1, 1, 1,...], indicating that it is the limit of a simple recursive pattern. This representation reveals the inherent symmetry and self-similarity of the golden ratio (Cohen, 2001).

In conclusion, transcendental numbers and continued fractions are essential components of advanced number patterns. Transcendental numbers, with their non-algebraic and non-rational nature, challenge our understanding of real numbers. Continued fractions, on the other hand, offer a powerful method for representing real numbers as infinite sequences of fractions, with the added benefit of uniqueness.

8.8. ADVANCED APPLICATIONS IN CRYPTOGRAPHY

In the ever-evolving landscape of cybersecurity, advanced applications in cryptography stand at the forefront of protecting sensitive information. Beyond the well-known RSA encryption, a wealth of cryptographic protocols and techniques leverage number theory to safeguard data in today's interconnected world. In this exploration, we delve into these advanced applications, the number-theoretic algorithms underpinning them, and the encryption techniques they employ (Caliceti et al., 2007).

8.8.1. Cryptographic Protocols Beyond RSA

Advanced cryptographic protocols have emerged to address evolving threats and requirements. One notable example is the Elliptic Curve Cryptography (ECC). ECC leverages the mathematical properties of elliptic curves over finite fields to provide robust security with shorter key lengths compared to RSA. Its security relies on the intractability of the elliptic curve discrete logarithm problem (ECDLP) (Sacristán et al., 2010).

Mathematically, ECC operates in the context of elliptic curves defined over finite fields, such as the equation:

$$y^2 = x^3 + ax + b \bmod p$$

Where (a), (b), and (p) are carefully chosen parameters. The security of ECC hinges on the difficulty of solving the ECDLP, which involves finding (k) when given (kP) and (P) where (P) is a point on the elliptic curve.

8.8.2. Number-Theoretic Algorithms and Their Security

Number theory plays a crucial role in the development of cryptographic algorithms. One prime example is the Diffie-Hellman key exchange protocol. This method enables two parties to securely establish a shared secret key over an unsecured communication channel. It relies on the computational complexity of the discrete logarithm problem, which is difficult to solve in large prime fields (Castillo, 2015).

In the Diffie-Hellman protocol, two parties agree on large prime numbers (p) and (g) (a primitive root modulo (p). Each party selects a private key (a) and (b), computes a public key (A) and (B), and then exchanges these public keys. The shared secret key is subsequently derived by both parties, which can be used for secure communication (Neville, 1971).

Additionally, cryptographic systems based on the hardness of integer factorization, such as the RSA cryptosystem, continue to be widely used. The security of RSA relies on the challenge of factoring the product of two large prime numbers, which remains computationally intensive.

8.8.3. Advanced Encryption Techniques Using Number Patterns

Advanced encryption techniques often leverage intricate number patterns to enhance security. One such method is the Rabin cryptosystem, which is based on the difficulty of factoring the product of two large prime numbers. It operates on the principle that encoding a message involves finding a square root modulo (n), where (n) is the product of two primes.

Mathematically, the Rabin encryption process involves representing a message as an integer (M) and encrypting it into a ciphertext (C) using the function $(C \equiv M^2 \bmod n)$. Decryption involves finding the four-square roots of (C) modulo (n) and then extracting the original message (Tanaev et al., 1994).

In conclusion, advanced applications in cryptography are a testament to the indispensable role of number theory in modern cybersecurity. Cryptographic protocols beyond RSA, number-theoretic algorithms, and advanced encryption techniques all rely on the mathematical foundation provided by number patterns. These techniques continue to evolve to meet the demands of an increasingly interconnected and data-driven world, where the security of sensitive information remains paramount.

8.9. QUANTUM COMPUTING AND NUMBER PATTERNS

Quantum computing stands as a revolutionary paradigm in computation, harnessing the principles of quantum mechanics to manipulate data in ways that classical computers cannot. At its core, quantum computing relies on quantum bits or qubits, which, unlike classical bits, can exist in superpositions of states, enabling the processing of vast amounts of information simultaneously (Raichev, 2012).

In the realm of number patterns, quantum computing promises to disrupt traditional cryptographic methods that rely on the difficulty of factorizing large numbers, a task that becomes trivial for quantum computers using Shor's algorithm.

- **Shor's Algorithm and Its Impact on Number Factorization:** Shor's algorithm, developed by mathematician Peter Shor in 1994, is a prime example of how quantum computing can be leveraged to crack one of the foundational problems in number theory – integer factorization. The algorithm's implications for cryptography are profound, as it dramatically reduces the time required to factorize large semiprime numbers, the cornerstone of RSA encryption.

Mathematically, Shor's algorithm exploits quantum Fourier transforms and modular exponentiation to efficiently determine the prime factors of a semiprime number N. In a classical context, factorizing large numbers is a time-consuming task, but Shor's algorithm has the potential to perform this task exponentially faster, posing a significant threat to classical cryptographic systems.

- **The Future of Number Patterns in Quantum Computing:** The advent of quantum computing introduces new possibilities and challenges for the field of number theory. On the one hand, the ability to factorize large numbers efficiently may necessitate the development of novel cryptographic techniques that are resistant to quantum attacks. Post-quantum cryptography seeks to address this concern by exploring cryptographic schemes that remain secure even in the era of quantum computing.

On the other hand, quantum computing offers exciting prospects for exploring number patterns and solving complex mathematical problems. Quantum computers excel at solving problems in quantum physics, optimization, and simulation that involve intricate mathematical structures. As researchers continue to harness the power of quantum computing, they may uncover new number patterns and mathematical relationships previously beyond reach (Kucukoglu & Simsek, 2018).

In conclusion, quantum computing is poised to reshape the landscape of number patterns and their applications. Shor's algorithm represents a significant advancement in the field of number theory, with profound implications for cryptography. The future of number patterns in the quantum era holds both challenges and opportunities as mathematicians and computer scientists work together to unlock the full potential of this groundbreaking technology (Qin et al., 2020).

Activity 8.1

In cryptography, advanced encryption techniques often rely on number patterns. Imagine you're a cryptographer tasked with designing a secure communication protocol that incorporates number patterns. How would you leverage the properties of Lucas numbers, Farey sequences, or continued fractions to enhance the security of your encryption system? Describe your approach and the benefits it offers.

8.10. SUMMARY

This chapter delves into the realm of advanced number patterns, exploring diverse mathematical concepts and their applications. It introduces the intricacies of Lucas number pentatopes, revealing their properties and close ties to Fibonacci numbers.

Farey sequences are discussed, offering insights into number theory and their applications in solving Diophantine equations. The chapter proceeds to elucidate Möbius transformations, showcasing their significance in conformal mapping and geometry. Partition theory and integer compositions are explored, emphasizing their role in combinatorics and number theory. Chaos theory and fractals are introduced, demonstrating their emergence in number sequences. The concept of transcendental numbers is defined with continued fractions as a representation method. Lastly, advanced applications in cryptography, including protocols beyond RSA, number-theoretic algorithms, and encryption techniques, are highlighted, along with the potential influence of quantum computing on number patterns. This chapter serves as a gateway to the more intricate and specialized aspects of number patterns, providing a deeper understanding of their significance in various mathematical disciplines, from number theory to cryptography and even quantum computing.

REFERENCES

1. Altinay, L., & Kozak, M., (2021). Revisiting destination competitiveness through chaos theory: The butterfly competitiveness model. *Journal of Hospitality and Tourism Management, 49*, 331–340.
2. Annamalai, C., (2022). Combinatorial and multinomial coefficients and its computing techniques for machine learning and cybersecurity. *The Journal of Engineering and Exact Sciences, 8*(8), 14713–01i.
3. Athanasiadis, D. I., Anton, N. E., Karim, A., Colgate, C. L., & Stefanidis, D., (2021). Does the advanced training in laparoscopic suturing enhance laparoscopic suturing skill beyond fundamentals of laparoscopic surgery? *Surgery, 170*(4), 1125–1130.
4. Bateman, P. T., & Horn, R. A., (1962). A heuristic asymptotic formula concerning the distribution of prime numbers. *Mathematics of Computation, 16*(79), 363–367.
5. Benavoli, A., Chisci, L., & Farina, A., (2009). Fibonacci sequence, golden section, Kalman filter and optimal control. *Signal Processing, 89*(8), 1483–1488.
6. Boeing, G., (2016). Visual analysis of nonlinear dynamical systems: Chaos, fractals, self-similarity and the limits of prediction. *Systems, 4*(4), 37.
7. Bohez, E. L. J., (1998). Two level cluster analysis based on fractal dimension and iterated function systems (IFS) for speech signal recognition. In: *IEEE. APCCAS 1998; 1998 IEEE Asia-Pacific Conference on Circuits and Systems. Microelectronics and Integrating Systems. Proceedings (Cat. No. 98EX242)* (Vol. 2, pp. 291–294). IEEE.
8. Caliceti, E., Meyer-Hermann, M. X., Ribeca, P., Surzhykov, A. X., & Jentschura, U. D., (2007). From useful algorithms for slowly convergent series to physical predictions based on divergent perturbative expansions. *Physics Reports, 446*(1–3), 1–96.
9. Carsault, T., Nika, J., Esling, P., & Assayag, G., (2021). Combining real-time extraction and prediction of musical chord progressions for creative applications. *Electronics, 10*(21), 2634.
10. Castillo, R. C., (2015). Recurrence relations and generating functions of the sequence of sums of corresponding factorials and triangular numbers. *Asia Pacific Journal of Multidisciplinary Research, 3*(4),

165–169.

11. Cohen, D. E., (2001). "The imperfect seeks its perfection": Harmonic progression, directed motion, and Aristotelian physics. *Music Theory Spectrum, 23*(2), 139–169.
12. De Weger, B. M., (1997). Equal binomial coefficients: Some elementary considerations. *Journal of Number Theory, 63*(2), 373–386.
13. Diamond, H. G., (1982). Elementary methods in the study of the distribution of prime numbers. *Bulletin of the American Mathematical Society, 7*(3), 553–589.
14. Edwards, H. M., (1996). *Fermat's Last Theorem: A Genetic Introduction to Algebraic Number Theory* (Vol. 50, pp. 875–880). Springer Science & Business Media.
15. Eger, S., (2013). Restricted weighted integer compositions and extended binomial coefficients. *J. Integer. Seq., 16*(13), 3.
16. Ellis, A. B., (2007). A taxonomy for categorizing generalizations: Generalizing actions and reflection generalizations. *The Journal of the Learning Sciences, 16*(2), 221–262.
17. Gaeini, A., Mirghadri, A., Jandaghi, G., & Keshavarzi, B., (2016). Comparing some pseudo-random number generators and cryptography algorithms using a general evaluation pattern. *IJ Information Technology and Computer Science, 9*, 25–31.
18. Garmany, B., & Müller, T., (2013). PRIME: Private RSA infrastructure for memory-less encryption. In: *Proceedings of the 29th Annual Computer Security Applications Conference* (Vol. 1, pp. 149–158).
19. Granville, A., (1997). Arithmetic properties of binomial coefficients. I. Binomial coefficients modulo prime powers. *Organic Mathematics (Burnaby, BC, 1995), 20*, 253–276.
20. Hassan, M. U., Rehmani, M. H., & Chen, J., (2019). Privacy preservation in blockchain based IoT systems: Integration issues, prospects, challenges, and future research directions. *Future Generation Computer Systems, 97*, 512–529.
21. Hilton, P., & Pedersen, J., (1989). Extending the binomial coefficients to preserve symmetry and pattern. *Computers & Mathematics with Applications, 17*(1–3), 89–102.
22. Hoggatt, Jr. V. E., (1967). Fibonacci numbers and generalized binomial coefficients. *Fibonacci Quart., 5*, 383–400.

23. Kalaivani, P., (2021). An advanced and shared cryptographic technique for security enrichment in ad hoc networks and mobile application. *Turkish Online Journal of Qualitative Inquiry, 12*(3), 657–660.
24. Karatsuba, A. A., (1990). The distribution of prime numbers. *Russian Mathematical Surveys, 45*(5), 99.
25. Kashdan, T. B., Gallagher, M. W., Silvia, P. J., Winterstein, B. P., Breen, W. E., Terhar, D., & Steger, M. F., (2009). The curiosity and exploration inventory-II: Development, factor structure, and psychometrics. *Journal of Research in Personality, 43*(6), 987–998.
26. Kazlacheva, Z. I., (2017). An investigation of application of the golden ratio and Fibonacci sequence in fashion design and pattern making. In: *IOP Conference Series: Materials Science and Engineering* (Vol. 254, pp. 1720–1723). IOP Publishing.
27. Konvalina, J., (2000). A unified interpretation of the binomial coefficients, the Stirling numbers, and the Gaussian coefficients. *The American Mathematical Monthly, 107*(10), 901–910.
28. Krcadinac, V., (2006). A new generalization of the golden ratio. *Fibonacci Quarterly, 44*(4), 335.
29. Kucukoglu, I., & Simsek, Y., (2018). Identities and derivative formulas for the combinatorial and Apostol-Euler type numbers by their generating functions. *Filomat, 32*(20), 6879–6891.
30. Le, T. K., Pham, P. V., Dong, C. L., Bahlawane, N., Vernardou, D., Mjejri, I., & Kim, S. W., (2022). Recent advances in vanadium pentoxide (V_2O_5) towards related applications in chromogenics and beyond: Fundamentals, progress, and perspectives. *Journal of Materials Chemistry C, 10*(11), 4019–4071.
31. Lee, J. S., & Chang, K. S., (1996). Applications of chaos and fractals in process systems engineering. *Journal of Process Control, 6*(2, 3), 71–87.
32. Lei, T., (1990). Similarity between the Mandelbrot set and Julia sets. *Communications in Mathematical Physics, 134*, 587–617.
33. Leroy, J., Rigo, M., & Stipulanti, M., (2016). Generalized Pascal triangle for binomial coefficients of words. *Advances in Applied Mathematics, 80*, 24–47.
34. Lin, X., Hofström, B., Wang, Y. P. E., Masini, G., Maattanen, H. L., Rydén, H., & Khan, T., (2021). 5G new radio evolution meets satellite communications: Opportunities, challenges, and solutions. *5G and Beyond: Fundamentals and Standards, 10*(3), 517–531.

35. Liu, Z., Choo, K. K. R., & Grossschadl, J., (2018). Securing edge devices in the post-quantum internet of things using lattice-based cryptography. *IEEE Communications Magazine, 56*(2), 158–162.

36. Losa, G. A., & Nonnenmacher, T. F., (1996). Self-similarity and fractal irregularity in pathologic tissues. *Modern Pathology: An Official Journal of the United States and Canadian Academy of Pathology, Inc., 9*(3), 174–182.

37. Mittal, S., Garg, S., Bhandari, H., & Sharma, V., (2022). A review on recent progressions of bismuth ferrite modified morphologies as an effective photocatalyst to curb water and air pollution. *Inorganic Chemistry Communications, 10*, 9834.

38. Mugford, G., (1999). Beyond basics: Giving greetings proper status in the EFL classroom. *MEXTESOL Journal, 22*(4), 11–20.

39. Murphy, J. G., Gex, K. S., Dennhardt, A. A., Miller, A. P., O'Neill, S. E., & Borsari, B., (2022). Beyond basics: A scoping review of novel intervention content to enhance the efficacy of brief alcohol interventions for emerging adults. *Psychology of Addictive Behaviors* (Vol. 2, pp. 258–260).

40. Neville, D., (1971). A technique for solving recurrence relations approximately and its application to the 3-J and 6-J symbols. *Journal of Mathematical Physics, 12*(12), 2438–2453.

41. Panda, M., (2014). Security in wireless sensor networks using cryptographic techniques. *American Journal of Engineering Research (AJER), 3*(1), 50–56.

42. Pathak, A. K., (2018). A simple probabilistic proof for the alternating convolution of the central binomial coefficients. *The American Statistician, 72*(3), 287–288.

43. Pham, Q. V., Fang, F., Ha, V. N., Piran, M. J., Le, M., Le, L. B., & Ding, Z., (2020). A survey of multi-access edge computing in 5G and beyond: Fundamentals, technology integration, and state-of-the-art. *IEEE Access, 8*, 116974–117017.

44. Phoenix, C., & Orr, N., (2017). Analyzing exceptions within qualitative data: Promoting analytical diversity to advance knowledge of ageing and physical activity. *Qualitative Research in Sport, Exercise, and Health, 9*(3), 271–284.

45. Qin, L., Cao, J., Shao, K., Tong, F., Yang, Z., Lei, T., & Peppas, N. A., (2020). A tumor-to-lymph procedure navigated versatile gel system

for combinatorial therapy against tumor recurrence and metastasis. *Science Advances, 6*(36), 3116.

46. Raichev, A., (2012). New software for computing asymptotics of multivariate generating functions. *ACM Communications in Computer Algebra, 45*(3, 4), 183–185.
47. Sacristán, A. I., Calder, N., Rojano, T., Santos-Trigo, M., Friedlander, A., Meissner, H., & Perrusquía, E., (2010). The influence and shaping of digital technologies on the learning–and learning trajectories–of mathematical concepts. *Mathematics Education and Technology-Rethinking the Terrain: The 17th ICMI Study, 276*(3), 179–226.
48. Sagan, B. E., (2020). *Combinatorics: The Art of Counting* (Vol. 210, pp. 547–550). American Mathematical Soc.
49. Schielack, V. P., (1987). The Fibonacci sequence and the golden ratio. *The Mathematics Teacher, 80*(5), 357–358.
50. Shparlinski, I. E., (2008). Arithmetic and geometric progressions in product sets over finite fields. *Bulletin of the Australian Mathematical Society, 78*(3), 357–364.
51. Son, J. B., (2014). *Moving Beyond Basics: From Call Coursework to Classroom Practice and Professional Development, 74*, 9872–9875.
52. Soundararajan, K., (2007). Small gaps between prime numbers: The work of Goldston-Pintz-Yildirim. *Bulletin of the American Mathematical Society, 44*(1), 1–18.
53. Spivey, M. Z., (2014). A combinatorial proof for the alternating convolution of the central binomial coefficients. *The American Mathematical Monthly, 121*(6), 537–540.
54. Srinivasan, T. P., (1992). Fibonacci sequence, golden ratio, and a network of resistors. *American Journal of Physics, 60*(5), 461–462.
55. Tanaev, V. S., Gordon, V. S., Shafransky, Y. M., Tanaev, V. S., Gordon, V. S., & Shafransky, Y. M., (1994). Priority-generating functions ordered sets of jobs. *Scheduling Theory – Single-Stage Systems, 5*, 186–252.
56. Van, G. R., (2014). The Fibonacci sequence and the golden ratio in music. *Notes on Number Theory and Discrete Mathematics, 20*(1), 72–77.

57. Wolf, M., (1997). $1/f$ noise in the distribution of prime numbers. *Physica A: Statistical Mechanics and its Applications, 241*(3, 4), 493–499.
58. Zang, X., Iqbal, S., Zhu, Y., Liu, X., & Zhao, J., (2016). Applications of chaotic dynamics in robotics. *International Journal of Advanced Robotic Systems, 13*(2), 60.

INDEX

K

L

M

N

O

P